iCons of matHematiCs

Edited by

G. Usha

Shanlax Publications
Madurai – 625003
shanlaxpublications@gmail.com
web: www.shanlaxpublications.com

ICONS OF MATHEMATICS

ISBN: 978-93-85977-34-3

Publisher

SHANLAX PUBLICATIONS
61, 66 T.P.K. Main Road, Vasantha Nagar
Madurai - 625003
Ph. 0452 – 4208765, 9600303383
shanlaxpublications@gmail.com
www.shanlaxpublications.com

Printed at

SHANLAX PRESS

61, 66 T.P.K. Main Road, Vasantha Nagar
Madurai - 625 003
Ph. 0452 – 4208765, 9942266683
email: shanlaxpress@gmail.com
web: www.shanlaxpress.com

DEDICATION

"A teacher takes a ***hand***, opens a ***mind*** and touches a ***heart***"

To Ms. Mayma Corera who has Academic Excellence, Vitality and a Flair for moulding creative minds.

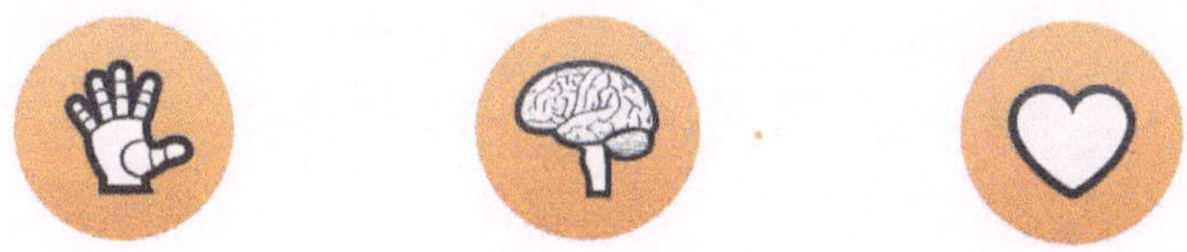

PSYCHOMOTOR COGNITIVE AFFECTIVE

PREFACE

Creation of nature and Mathematics are closely related. The curiosity of human being towards nature might be considered as the source of spring of Mathematics. Questions like what? how? why? and how many? might have arisen within him by seeing the environment and sky, hearing the sound, touching the objects etc. In order to get solution to these questions the primitive human would have to make an approach of logical reasoning and rational thinking. This logical approach and rational thinking are bases of Mathematics.

"Just like the work of a painter, who makes out patterns with his paints, a mathematician works with nothing but his ideas and his ideas similar to a painter's strokes must be beautiful" says the famous mathematician Bertrand Russell. Only then can he produce a work that can be called profoundly significant and creative. Mathematics is often defined as a study of pattern and so it is.

Mathematics is a field where many people feel diffident, but there are some who have a passion for numbers and making discoveries regarding equations, measurements, and other numerical solutions in history. Did all these great mathematicians of the world have anything in common? Are they born savants? Did they realize the existence of theorems and other mathematical discoveries in their journey of life? There are two schools of thoughts that answer these questions. Plato believed that these theorems and concepts are out there in the universe just waiting to be revealed. But the 'realists' have a different philosophy; according to them mathematical concepts and ideas arise from our physical presence and we solely are responsible for their being. Whichever theory is true the fact remains that Mathematics is creative boldness and all these famous mathematicians found that creativity and boldness within them that encouraged them to discover the mathematical principles we now know and study in our books.

Often called the language of the universe, Mathematics is fundamental to our understanding of the world. Everywhere we look it is likely Mathematics has made an impact, from the faucet in our kitchen to the satellite that beams our television programs to our home. As such, great mathematicians are undoubtedly risen and going to rise above the rest and have their names embedded within history. This book documents some such people. I hope one looks deeper into the lives of these men, as they are truly fascinating people and their discoveries are astonishing – too much to include here. Have you ever wondered how these magical mathematical concepts have arrived and who are the people behind their invention?

You can read through the book and learn more about these people and how they used their mathematical abilities to make their mark on the world.

Every great athlete, artist and aspiring being has a great team to help them flourish and succeed - personally and professionally. A great many people have contributed to its production. I owe my sincere gratitude to all those people who have made this book possible.

To Dr. Josephine Salathe Mary - Thank you for your expertise, time, and feedback. You have a knack for providing corrective feedback in the most positive way.

To all the First Year student-teachers of Mathematics – Thank you all for your contribution. From the daily words of encouragement to the covering page, when my time and attention were focused on this, you truly are an amazing group of individuals to work with.

Most importantly, none of this would have been possible without the love, patience, and support of my family.

There are two ways to do great Mathematics. The first is to be smarter than everybody else. The second way is to be stupider than everybody else—but persistent.

Take the first step and your mind will mobilize all its forces to your aid. But the first essential is that you begin. This is just a beginning for us.

G. Usha

CONTENT

Usha G
Asst. Prof. in Mathematics

Chapter 1

Srinivasa Ramanujan

Indian Government issued postage stamps to celebrate his 75th birthday.

Born : **December 22, 1887, Erode**
Died : **April 26, 1920**
Nationality : **Indian**

Srinivasa Ramanujan, an Indian mathematician was born on December 22, 1887, in Erode, India, a small village in the southern part of the country. Shortly after this birth, his family moved to Kumbakonam, where his father worked as a clerk in a cloth shop. He first went to school at five and at the age of seven he was transferred to town high school at Kumbakonam, where he held a scholarship. He was quiet meditative and had strong memory.

After demonstrating an intuitive grasp of Mathematics at a young age, he began to develop his own theories and in 1911 published his first paper in India. Two years later Ramanujan began a correspondence with British mathematician G. H. Hardy that resulted in a five-year-long mentorship for Ramanujan at Cambridge, where he published numerous papers on his work and received a B.S. for research. His early work focused on infinite series and integrals, which extended into the remainder of his career.

Contributions of Ramanujan to Mathematics

- His greatest contribution to Mathematics was in the field of numbers.
- In the simplest arrangement of digits he found wonderful properties, consequences, symmetries and relationship that had escaped the notice of even the ablest of mathematicians up to that time.
- He studied the highly composite numbers also which are recognized on the opposite of prime numbers.
- In 1913, he wrote original theorems in elegant English.
- One day he asked his Mathematics teacher, about zero and infinity. He questioned, zero/zero is one. His classmates laughed at his question. Then the teacher explained other students. Some mathematicians claimed it was unity.
- Indian mathematician Bhaskara proved that it was infinity. So the teacher appreciated his prodigy.
- At the age of 13, he got Loney's Trigonometry from a college library; he started his own research with theorems.
- At the age of 15, his senior friends gave him synopsis of elementary results in pure and applied Mathematics by George Carr. He used to write his ideas and results on loose sheets.
- He set out 120 theorems and formulae about Reimann series, a topic in the definite integral calculus, to the Hardy.
- During an illness in England, Hardy visited Ramanujan in Hospital. He had taken taxi number 1729. Ramanujan said, it is the smallest integer that can be represented in two ways by the sum of two cubes.

$$1729 = 1^3 + 12^3 = 9^3 + 10^3$$

- The elementary analysis of highly composite numbers is most remarkable and it shows very clearly Ramanujan's extraordinary mastery over the algebra of inequalities.
- Ramanujan derived pleasure from the game of fiddling with numbers. He presented many starling statements about numbers. For example he made the statement that every even integer greater than two is the sum of two primes. (i.e) numbers having divisions. Thus

$$4 \longrightarrow 2+2$$
$$6 \longrightarrow 3+3$$
$$8 \longrightarrow 3+5$$

- Partition of whole numbers is another similar problem which engaged the attention of Ramanujan. Take the case of number 3. There are 3 alternative ways to write it. (i.e) 3 + 0, 1 + 2, 1 + 1 + 1. There is no other way of partitioning this number. Similarly 4 in 5 ways. 4 + 0, 3 + 1, 2 + 2, 2 + 1+ 1, 1 + 1 + 1 + 1 if only integers are to be used.

- This process of partitioning is very difficult, but Ramanujan developed a formula and that is valid for any number. The formula is universally applicable whether 3 or 30000000.
- He also investigated divergent series.

Of Ramanujan's published papers - 37 in total - Professor Bruce C. Berndt reveals that "a huge portion of his work was left behind in three notebooks and a 'lost' notebook. These notebooks contain approximately 4000 claims, all without proofs. Most of these claims have now been proved, and like his published work, continue to inspire modern-day Mathematics."

Ramanujan wrote down all his mathematical findings, in ledger-like notebooks that lately became famous as Ramanujan's Frayed Notebooks. The Cambridge University Press brought out his Collected Papers in 1927 and after this, the mathematicians all over the world became fascinated by his work and personality. In 1929, two British mathematicians, G. N. Watson of the University of Birmingham and B. M. Wilson of Liverpool University, started to study and edit the notebooks with an intention to unearth the mathematical gems lying undiscovered in them. The Notebooks contained numerous formulas most of which were related to mock theta functions and q-series and around a third are about modular equations and singular module and other formulas of congruencies, integrals, and asymptotic series.

Eventually, the Tata Institute of Fundamental Research, Mumbai published the facsimile editions of the two volumes of Ramanujan's notebooks in 1957. The books were published at the initiative of the Indian nuclear physicist Dr. Homi Jehangir Bhabha and the publication once again, renewed the mathematicians' interest in Ramanujan's notebooks. The most outstanding contribution of Ramanujan was his formula for p (n), the number of partitions of "n". For all these reasons, he is hailed as an all-time great mathematician like Euler, Gauss or Jacobi for his natural intellect.

Srinivasa Ramanujan died on 26 April 1920, at the age of 32 at hands of dreadful disease of tuberculosis. And even on his deathbed he was writing down theorems that had come to him in a dream. These and many of his earlier theorems are so complex that the full scope of Ramanujan's legacy has yet to be completely revealed and his work remains the focus of much mathematical research. His collected papers were published by Cambridge University Press in 1927.

A biography of Ramanujan titled "The Man who knew Infinity" was published in 1991 and a movie of the same name starring Dev Patel as Ramanujan and Jeremy Irons as Hardy, premiered in September 2015 at the Toronto Film Festival.

Chapter 2

Shakuntala Devi

Human computer

Born	**: November 4, 1929, Bengaluru, Karnataka**
Died	**: April 21, 2013, Bengaluru, Karnataka**
Nationality	**: Indian**

India is a land popular for its contribution in the field of Mathematics. It has always been proud of its mathematical prodigies. One such prodigy in recent times of modern India was Shakuntala Devi who was world renowned and referred to as "Human Computer".

Early Life

Shakuntala Devi was born on 4th of November, 1929 in Bengaluru in a well-known Brahmin priest family. Born in a poor family in Bangalore (India), Shakuntala Devi dropped out of school because her father, a circus worker, could not afford the monthly school fee of Rs 2. She grew up in a slum and at a very young age, her mathematical abilities were recognized by her father. There is an interesting anecdote about how he recognized it.

At the age of 3, when she started playing cards with her father, he was surprised to find that she was winning all the games against him every day.

Suspecting some foul play, he began his "investigation" during which he realized that she was memorizing all the card numbers and their sequence as the game progressed in the initial rounds and with her memory power, she was able to predict the sequence of cards in the subsequent rounds in the same game and thus wait to pick cards strategically to help her win.

Her father taught her mathematical operations like multiplication, division and square root and took her to his circus to demonstrate her quick calculation abilities and memory power to the crowds. As the word about her skills spread, she started doing road shows as well across the city. At the age of 6, she gave her first major show at Mysore University and there was no turning back after that.

Recognition and Fame

- In her early 20s, she toured Europe extensively to demonstrate her skills. During an interview on BBC, she was given a complicated mathematical calculation which she solved within seconds but her answer was different from what the interviewer and his team had calculated. When she insisted that her answer was right, the interviewer and his team of math experts reexamined their calculations for several minutes and finally admitted that their initial calculations were wrong. That incident spread like wildfire across the world after which she was being popularly referred to as "The Human Computer". During her Europe tour, she was invited by several reputed Universities including the University of Rome.
- In 1977, at Southern Methodist University in Dallas, she calculated the 23rd root of a 201-digit number. It had taken four minutes for a professor to write the problem on the board, and it took more than a minute for a Univac computer to figure out the answer. Shakuntala Devi got it in 50 seconds. Here is a newspaper excerpt of the same:

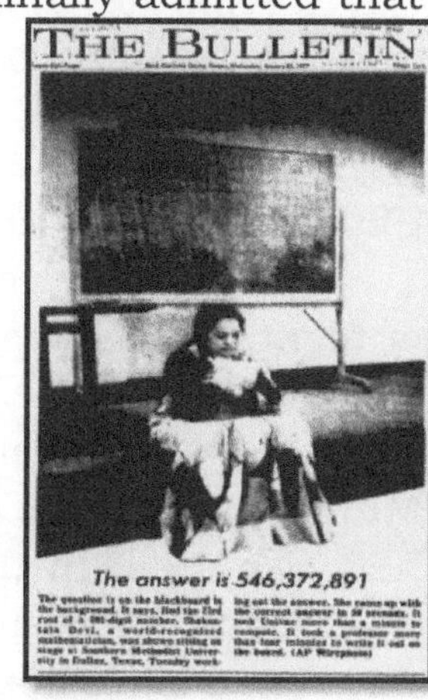
THE BULLETIN

The answer is 546,372,891

- After the 1977 record at Dallas, she went on to create more records including one in 1980 where she solved multiplication of two 13 digit randomly picked numbers in just 28 seconds and entered the Guinness Book of World Records.

Inspiration

Not only was she a genius and mathematical magician but also a role model for students and math enthusiasts. She played a great role in making Mathematics an interesting subject for millions of students due to her practical, fast and efficient approach towards problem solving. She had authored several books for children to help them develop interest in Maths and help them understand the subject better. Her books on puzzles are popular among college students preparing for jobs interviews which require analytical skills.

Her workshops on Maths techniques and Vedic Mathematics were popular throughout the nation. On the social front, she started "Shakuntala Devi Education Foundation Public Trust" with a mission to provide quality education for children of deprived sections of the society. Also, she encouraged research in Vedic Mathematics and helped spread global awareness about India's contribution towards Mathematics.

Awards and Achievements

- Shakuntala Devi won the 'Distinguished Woman of the Year Award' in 1969, from the University of Philippines along with a gold medal.
- In 1988, she was honoured with the 'Ramanujan Mathematical Genius Award' in Washington D.C., conferred to her by the-then Indian Ambassador to US.
- Her name was listed in the '1995 Guinness Book of World Records' edition for her outstanding mathematical feat where she beat the world's fastest computer at multiplying two thirteen digit numbers.
- A month before her death, she was honoured with the 'Lifetime Achievement Award' in Mumbai, in 2013.

Shakuntala Devi as an Author

Shakuntala Devi has authored a few books. She shares some of the methods of mental calculations in her world famous book, Figuring: The Joy of Numbers. Puzzles to puzzle You, More Puzzles to puzzle you, The Book of Numbers, Mathability: Awaken the Math Genius in Your Child, Astrology for you, Perfect Murder, In the Wonderland of Numbers are some of the popular books written by her. Her book, In the Wonderland of Numbers, talks about a girl Neha, and her fascination for numbers. Her other famous books include Astrology for You and The World of Homosexuals.

Shakuntala Devi is an excellent astrologer as well and provides remedies based on date and time of birth and place. Her clients include celebrities and well known personalities from various fields.

Death and legacy

On April 2013, Devi was admitted to a hospital in Bengaluru with respiratory problems. Over the following two weeks she suffered from complications of the heart and kidneys. She died in the hospital on 21st April 2013. She was 83 years old then. She is survived by her daughter, Anupama Banerji. On 4th November 2013, Devi was honoured with a Google Doodle for what would have been her 84th birthday.

She was a math evangelist, human computer, social worker and role model, and India is indeed proud to be home to such an inspirational personality.

Chapter 3

Aryabhata

Aryabhata is credited to have set up an observatory at the Sun temple in Taregana, Bihar.

Born : 476 A.D., Kusumapura, Taregana
Died : 550 A.D.
Nationality : Indian

Aryabhata was one of the first acclaimed Indian mathematicians and astronomers belonging to the classical age. He was born in Kusumapura (present day Patna) - Taregana a small town in Bihar, India. His contribution to Mathematics, science and astronomy is immense. Some archaeological evidence suggests that he came from the present day Kodungallur, the historical capital city of Thiruvanchikkulam of ancient Kerala - this theory is strengthened by the several commentaries on him having come from Kerala.

He went to Kusumapura for advanced studies and lived there for some time. Both Hindu and Buddhist traditions, as well as Bhaskara I, the 7th Century mathematician, identify Kusumapura as modern Patna. Aryabhata's work was of great influence in the Indian astronomical tradition and influenced several neighbouring cultures through translations. Some of his works are cited by Al-Khwarizmi, and in the 10th century by Al-Biruni. The Aryabhata Knowledge University (AKU), Patna, has been established by the Government of Bihar in his honour for the development and management of educational infrastructure related to technical, medical, management and allied professional education.

India's first satellite Aryabhata is named in his honour. At the Aryabhata Research Institute of Observational Sciences (ARIOS) near Nainital, India, research in astronomy, astro Physics and atmospheric sciences is conducted.

Contributions to Mathematics

- He wrote on arithmetic which included arithmetic and geometric progressions, quadratic equations and indeterminate equations.
- He also computed arithmetical tables which contained a table of sines.
- His work in mensuration was crude and in this respect it somewhat resembled that of the Egyptians. For example, he gave the area of a triangle by a scheme that applies only to the case of an isosceles triangle.
- He was also interested particularly in the series, permutations and linear and quadratic equations.
- He was familiar with the equations of the form $ax^2 + bx + c= 0$, but he did not tell how he approached their solution.
- His work often called the Aryabhata consists of a collection of astronomical tables, and the Aryastasta which includes the Ganita, a note on arithmetic, the Kalakriya on time and its measure and the Gola, on the sphere.
- The arithmetic carries numeration by tens as far as 10^{10}treats of plane and solid numbers, and gives a rule for square root.
- His work contains one of the earliest attempts at a general solution of a linear indeterminate equation by the method of continued fractions.
- He gave the volume of a pyramid as half of the product of the base and the height and the volume of a sphere as $\frac{3}{2}\Pi r^3$.
- He gives at one place an extremely accurate value of π viz. $3\ \frac{177}{1250}$ (=3.1416), but he himself never utilized it, nor did any other Hindu mathematician before the twelfth century.
- His rule about π is quite interesting --- "Add 4 to 100, multiply by 8, add 62000", the result is the approximate value of the circle when the diameter is 20000. This makes π equivalent to $\frac{6\ 28\ 32}{20000}$ or 3.1416.
- He was the one who suggested the use of letters to represent unknown.
- He gave a rule for the extraction of cube root.
- He also gave formulae for the area of a square, rectangle, triangle, rhombus and circle.
- He revealed the geometrical fact, "In a square the product of two Saras is the square of the half chord of the two arcs (i.e.) $a \times b = c^2$ where c is half the chord and the Saras (arrows) are the segments of the diameter which bisects any chord.
- He gave almost all formulae for knowing areas of different figures. He gave the formula $(a + b)^2 = a^2 + 2ab + b^2$

- He gave a rule for summing an arithmetic series after the pth term.
- Aryabhata held that the earth was a spherical body and that it revolved around the sun.
- He had also made calculations of the rotation of the earth on its axis.
- He also appears to have had the knowledge of gravitational pull of the earth.
- His approach was scientific. He also dealt with involution, area, circular arcs, algebraic equations, and identities.
- In his book Aryabhata had given 121 shlokas which have been divided into four parts (i) Gati-Paadika (ii) Ganit Paadika (iii) Kaal Kriyapad and (iv) Ga-Pad.
- His book has consolidated a very wide subject matter in a very small group.

Aryabhata was a very intelligent man. The theories that he came up with at that time present a wonder to the scientific world today. His works were used by the Greeks and the Arabs to develop further. A commentary by Bhaskara I a century later on Aryabhatiya says:

'Aryabhata is the master who, after reaching the furthest shores and plumbing the inmost depths of the sea of ultimate knowledge of Mathematics, kinematics and spherics, handed over the three sciences to the learned world.'

Aryabhata was an immense influence to Mathematics and astronomy. Many of his works inspired Arabs more particularly. His astronomical calculations helped form the 'Jalali calendar'. He has been honoured in many ways. The first Indian satellite is named after him as 'Aryabhata', so is the lunar crater. An Indian research center is called 'Aryabhata Research Institute of Observational Sciences'.

Chapter 4

Brahmagupta

Thank you
Brahmagupta for your
zero

Born : **598 A.D., Bhinmal, Rajasthan**
Died : **670 A.D.**
Nationality : **Indian**

Brahmagupta was an Indian mathematician, born in 598 A.D. in Bhinmal, a state of Rajasthan of northwest India (he is often referred to as Bhillamalacarya, the teacher from Bhillamala), and later became the head of the astronomical observatory at Ujjain in central India. Most of his works are composed in elliptic verse, a common practice in Indian Mathematics at the time, and consequently have something of a poetic ring to them.

Influenced by the spread of Greek mathematical ideas eastward during the imperial expansion of the ancient Roman empire, Brahmagupta's ideas in turn had an impact on later European developments; they were translated into Arabic from his own Sanskrit language, and thus took their place among the foundation stones of Western Mathematics. Brahmagupta's writings contain mathematical and astronomical concepts that are taken for granted today, but they were concepts that he pioneered or refined from ideas he inherited.

Contributions of Brahmagupta to Mathematics

- He gave the rules for negative numbers in his algebra, and solved quadratics.

- He assumed, for example, that the area of an isosceles triangle was half the product of the base and one of the equal sides.
- In writing fractions, the scheme of writing the numerator above the denominator was used by him.
- He was particularly concerned with series and permutations.
- His knowledge of quadratics is shown in his solving the equation by a rule that is equivalent to the quadratic formula.
- He gave the rules of sines for their operation and distinguished between the two classes of numbers by writing a dot or a small circle over those that were negative.
- He gave $\frac{22}{7}$ as the value of π and suggested 3 as a practical value.
- He contributed a lot in understanding the concept of zero.
- He defines it as a-a=0.
- He gave the rules for determining the sum of the squares and cubes of natural numbers.
- He gave four different methods of multiplication.
- He wrote an astronomical work in twenty-one chapters entitled Brahmasidhanta, when he was thirty years old.
- His arithmetic includes work with integers, fractions, progressions, barter, simple interest, mensuration of plane figures, and problems on volumes and on shadow reckoning.
- Area of quadrilateral whose sides are a,b,c,d, is

  $$A = \sqrt{(s-a)(s-b)(s-c)(s-d)} \quad \text{where}$$

  $$s = \frac{1}{2}(a+b+c+d)$$

  A formula that is true only for cyclic quadrilaterals.
- He has given the formula for the triangular area as

  $$\sqrt{(s-a)(s-b)(s-c)(s-d)}$$
- He gave the usual rules for negative numbers.
- In the case of simultaneous equations of the first degree the unknowns are spoken of as colours and the problems are chiefly astronomical.
- He wrote that the length of a year was 365 days 6 hours 12 minutes 9 seconds.
- He indicated that Earth was nearer the moon than the sun.

This great mathematician died between 660 and 670. He was one of the greatest mathematicians in Indian history and his contributions to Mathematics and science have made major differences to various mathematical problems by establishing the basic rules which now allow us to find their solutions. He was honoured by the title given to him by a fellow scientist 'Ganita Chakra Chudamani' which is translated as 'The gem of the circle of mathematicians'.

Chapter 5

Bhaskara

> Objects fall on earth due to a force of attraction by the earth. Therefore, the earth, planets, constellations, moon, and sun are held in orbit due to this attraction.

Born : 1114 C.E., Bijapur, Hyderabad
Died : 1185 C.E.
Nationality : Indian

Bhaskaracharya was born in 1114 C.E. at some place near Bijapur in Hyderabad. Bhaskaracharya's name was actually Bhaskara but the title 'Acharya' was added and conferred to mean "Bhaskara the Teacher". He is also known as Bhaskaracharya II. Bhaskaracharya's father named Mahesvara was a Brahmin of Shandilya Gotra, well versed in Shroud (originated from "shut" or "Vedas") and "Smart (originated from "smut") Dharma, respected by all and who was an authority in all the branches of knowledge. Bhaskaracharya acquired knowledge at his feet. We also come to know that Bhaskaracharya was a resident of Vijjadveed and that his father taught him Mathematics and Astronomy. Vijjadveed surrounded by the hills of Sahyadri was the centre of learning at the time of Bhaskaracharya.

The subjects of his six works are arithmetic, algebra, trigonometry, calculus, geometry, and astronomy. The six works are: Lilavati on Mathematics; Bijaganita on algebra; the Siddhantasiromani which is divided into two parts: mathematical astronomy and sphere; the Vasanabhasya of Mitaksara which is

Bhaskaracharya's views on the Siddhantasiromani; the Karanakutuhala or Brahmatulya in which he simplified the concepts of Siddhantasiromani and the Vivarana which comments on the Shishyadhividdhidatantra of Lalla.

From the mathematical point of view the first three of these works are the most interesting.

Contributions of Bhaskara to Mathematics

- ✓ Siddhanta Shiromani is divided into four parts and each part is subdivided into several chapters. The Siddhanta Shiromani (written in 1150) demonstrates Bhaskara's knowledge of trigonometry, including the sine table and relationships between different trigonometric functions. He also discovered spherical trigonometry, along with other interesting trigonometrical results. In particular Bhaskara seemed more interested in trigonometry for its own sake than his predecessors who saw it only as a tool for calculation. Among the many interesting results given by Bhaskara, discoveries first found in his works include the now well known results for sin (a+b) and sin (a-b) :
  - $\sin(a+b) = \sin(a)\cos(b) + \cos(a)\sin(b)$
  - $\sin(a-b) = \sin(a)\cos(b) - \cos(a)\sin(b)$
- ✓ The first part is known as 'Lilavathi'. It is said that Lilavathi was the name of Bhaskara's daughter and that astrologers predicted that she should never wed. Baskara, however, divined a lucky moment for her marriage and left an hour-cup floating on the vessel of water. This cup had a small hole in the bottom and was so arranged that the water would trickle in and sink it at the end of the hour. Lilavathi with a natural curiosity looked in the cup and a pearl from her garments chanced to drop in the cup and stop the influx of water. So the hour passed without the sinking of the cup and Lilavathi was thus fated. To console her, Bhaskaracharya wrote a book after her name.
- ✓ Lilavathi includes notation, the operations with integers and fractions, the rule of three the most common commercial ruler, interest series, allegation, permutation, menstruation and a little algebra.
- ✓ The ruler of zero is also given. In this book questions have been composed in a very interesting manner.
- ✓ He introduced for the first time, the idea of infinity while dividing a number by zero.
- ✓ He also gave many important formulae for measurements.

  (Eg.) Area of sphere = 4 x Area of circle.

  Volume of sphere = Area of sphere x ⅙ of its diameter.
- ✓ He also gave the formulae for permutation and combination

  (Eg.) No. of permutations of r things = └t / └k └l where k and l are different objects.
- ✓ He also has the knowledge of surds.
- ✓ His book "Big Ganit" is a book on Algebra.

- ✓ He also dealt with cubic and bi - quadratic equations such as

  $x^3 + 42x = 6x^2 - 65.$

  $x^4 - 3x^2 - 900x = 9999.$
- ✓ In this, he has discussed directed numbers, the negatives being designated as debt of loss and being indicated by a dot over each number and the usual rules and formulae being stated correctly. He dismissed the imaginary by stating:

  "There is no square root of negative quantity for it is not a square".
- ✓ He presented a complete and systematic explanation of the Indian method of solving determinate and indeterminate problem on various topics including sale and purchase, interest, gems, gold, permutations of syllables in a stanza, military marches, excavations and measure of grains.
- ✓ He also gave a proof of Pythagoras theorem. He gave a method of deducing new sets of solutions of $x^2 + 1 = y^2$. He also discussed $x^2 + B = y^2.C$ where B and C are non-square integers.
- ✓ He presented complicated and abstract theorems in poetic form.
- ✓ He had the knowledge of the gravitational power before Newton.
- ✓ His method for finding the solutions of the problem $x^2 - ny^2 = 1$ (so-called "Pell's equation") is of considerable interest and importance.
- ✓ He also gave solutions of Diophantine equations of the second order, such as $61x^2 + 1 = y^2$. This very equation was posed as a problem in 1657 by the French mathematician Pierre de Fermat, but its solution was unknown in Europe until the time of Euler in the eighteenth century.
- ✓ He contributed to the preliminary concept of infinitesimal calculus, along with notable contributions towards integral calculus.
- ✓ He stated Rolle's theorem, a special case of one of the most important theorems in analysis, the mean value theorem. Traces of the general mean value theorem are also found in his works.

Lilavati

Lilavati was the daughter of great ancient Mathematician and astrologer, Bhaskaracharya. Bhaskara was born in the Sahyadri hills, at present situated in Bijapur, Mysore. Teaching and learning Mathematics was an ancestral vocation with her. He learnt Mathematics from his father. About Lilavati, story is prevalent that Bhaskara's daughter's name was Lilavati after whom he named the work. A good astrologer that he was, Bhaskara came to know from her horoscope that her married life would be cut short. Bhaskara computed that the disaster of her married life can be averted if her marriage was solemnized on a definite date punctually at a specified time. He made all arrangements for this.

For measuring time he constructed a sand bowl in which sand would flow from one vessel to another beneath it through a small orifice at fixed intervals of time. On the day previous to the marriage, Lilavati inquisitively looked into the new instrument which had been installed in her house and while looking a small pearl from the ornament on her nose fell into the sand and got mixed with it. This retarded the movement of the sand with the result that the marriage was celebrated later than the time which has been fixed after careful astrological calculations. Lilavati lost her life. To console her in her lifelong grief Bhaskara taught her arithmetic and named his work after her.

The Lilavati contains 278 verses and deals with the various subjects e.g. Units of weights and measures, 20 arithmetic operations viz, addition, subtraction, multiplication, division, square root, cube, cube roots, five rules of reduction of fraction of rule of 3, 5, 7, 9, 11 terms and barter, 8 arithmetical determinations, mixture, progression, plane figure, excavation, stock, saw mounds of grain and shadow of a gun man, indeterminate equations of second degree, triangles, quadrilaterals, areas of circle, volumes of spheres, cone, pyramids, etc. A number of commentaries on the translation of Lilavati have appeared within the last eight centuries.

Some scholars have suggested that Bhaskara's work influenced later developments in the Middle East and Europe. His work was perhaps known to Islamic mathematicians as soon as it was written, and influenced their subsequent writings. The results thus became indirectly known in Europe by the end of the twelfth century, but the text itself was not introduced until much later (Ball, 1960). The Mughal emperor Akbar commissioned a famous Persian translation of the Lilavati in 1587.

Bhaskara was a natural born teacher and mathematician. As is common at his time, generations of a family would be mathematicians, with each father passing on knowledge to their sons. Bhaskara himself passed on his knowledge to his son, Loksamudra. Bhaskaracharya passed away in 1185 C.E. at Ujjain. In a journal of Royal society, Dr. Stopwood has remarked “We must acknowledge the intimate capabilities of Bhaskaracharya”. His contribution to Astronomy and Mathematics in the 12th century was unparalleled and significant.

Chapter 6

Euler

He produced on average one mathematical paper every week in the year 1775.

Born : **April 15, 1707**
Died : **September 18, 1783**
Nationality : **Swiz**

Leonhard Euler was born on April 15, 1707, in Basel, Switzerland, to Paul Euler and Marguerite Brucker in Basel. His father was a pastor of the Reformed Church while his mother was the daughter of a pastor, thus Euler was raised in a religious environment. He had two sisters. His father taught the little boy Mathematics, kindling in him a lifelong love for the subject. Leonhard's father was friends with Johann Bernoulli, a prominent mathematician who would greatly influence the young boy.

At an early age in the University of Boset, he attracted the attraction of John Bernoulli. He received the degree of Master of Arts at the age of 17 wherein he submitted a dissertation on “Comparison between Cartesian and Newtonian system”. At the age of 19, he transmitted two dissertations to Paris Academy. One of them was on Masting of Ships and the other on philosophy of sound. In 1733, he married a Swiss lady the daughter of a painter who had been brought to Russia by Peter the great. In 1735, he found the solution of the problem in three days which otherwise deemed solvable in several months.

At the age of 30 in 1737, he was honoured by Paris Academy for dissertation upon the efflux and reflux of the sea. In 1741, he was invited by King Fredrick the Great of Germany to live there. He remained there till 1766. Here he prepared a set of lessons on nature philosophy and a model of interesting teaching for the princess of Anhalt Dessau. It is noteworthy that he could find time for such elementary works amidst his other literary interests. A fire broke out in his house in 1771 in which all his books were burnt but his writings were safe.

Contributions of Euler to Mathematics

- He was a prolific author and had written numerous articles and textbooks. His textbook 'Mechanica' was published in 1736–37 in which he presented Newtonian dynamics in the form of mathematical analysis.
- Another textbook, 'Introductio in analysin infinitorum' was out in 1748 in which he developed the concept of function in mathematical analysis. His works were of great significance to the field of modern analytic geometry and trigonometry.
- In addition to Mathematics he was also interested in astronomy and worked on developing a theory of lunar motion which involved the interactions of Sun, Moon, and Earth. He could only devise a partial solution which was published in 1753.
- He wrote a textbook in calculus, 'Institutiones calculi differentialis' in 1755 and another one, 'Institutiones calculi integralis' during the period 1768–70. These works served as the foundation for modern calculus as they contained formulas for integration and differentiation.
- He explained the basic principles of mechanics, optics, acoustics, and astronomy in his 'Lettres à une princesse d'Allemagne' which was published from 1768 to 1772.
- He proved Newton's identities, and also several of the theories propounded by the mathematician, Fermat, including Fermat's little theorem and Fermat's theorem on sums of two squares.
- He worked on Fermat's little theorem and using his own findings and the functions he had developed, he gave the Euler's theorem, which is in effect a generalized version of Fermat's little theorem.
- His works played a key role in the development of the Euler-Bernoulli beam equation which is today considered a cornerstone of engineering. He was also very popular for using his scientific findings to solve real-world problems.
- Euler is counted among the greatest mathematicians to have ever walked this earth. He had made tremendous contributions to the field of Mathematics and is the only mathematician to have had two numbers named after him: Euler's Number in calculus, e, approximately equal to

2.71828, and the Euler–Mascheroni constant γ (gamma), approximately equal to 0.57721.

- Euler introduced several notational conventions. He was the first to write f(x) to denote the function, the Greek letter Σ for summations, and the letter 'e' for the base of the natural logarithm.
- His ultimate aim was the perfection of analysis and calculus. In connection with one simple formula e^2 x + 1=0. Folix Klein has remarked, "All analysis was centered here."
- Every symbol has its history - the principal whole number 0 and 1;
- The chief mathematical relations – and =; the sign for 'impossible' square root of minus over 'e' the base of Nigerians logarithm.
- Euler made important contributions to complex analysis. He introduced the scientific notation. He discovered what is now known as Euler's formula, that for any real number φ , the complex exponential function satisfies
- $e^{i\varphi} = \cos\varphi + i\sin\varphi$.
- This has been called "the most remarkable formula in Mathematics" by Richard Feynman. Euler's identity is a special case of this:
- $e^{i\pi} + 1 = 0$.
- This identity is particularly remarkable as it involves *e*, π, *i*, 1, and 0, arguably the five most important constants in Mathematics.
- Euler is well known in analysis for his frequent use and development of power series: that is, the expression of functions as sums of infinitely many terms, such as

$$e = \sum_{n=0}^{\infty} \frac{1}{n!} = \lim_{n\to\infty}\left(\frac{1}{0!}+\frac{1}{1!}+\frac{1}{2!}+\ldots+\frac{1}{n!}\right)$$

- Notably, Euler discovered the power series expansions for e and the inverse tangent function

$$\arctan z = \sum_{n=0}^{\infty} \frac{\left(-1^{n} z^{2n+1}\right)}{2n+1}$$

- In 1736 Euler solved, or rather proved unsolvable, a problem known as the seven bridges of Konigsberg. . Euler's solution of the Konigsberg bridge problem is considered to be the first theorem of graph theory. In addition, his recognition that the key information was the number of bridges and the list of their endpoints (rather than their exact positions) presaged the development of topology.
- He also facilitated the use of differential equations, in particular introducing the Euler–Mascheroni constant:

$$\gamma = \lim_{n\to\infty}\left(1+\frac{1}{2}+\frac{1}{3}+\frac{1}{4}+\ldots+\frac{1}{n}-\ln(n)\right).$$

- Leonhard Euler is also recognized for using closed curves to demonstrate syllogistic reasoning (1768). These illustrations are known as Euler diagrams.

- Despite a long life and thirteen children, Euler had more than his fair share of tragedies and deaths, and even his blindness later in life did not slow his prodigious output - his collected works comprise nearly 900 books and, in the year 1775, he is said to have produced on average one mathematical paper every week - as he compensated for it with his mental calculation skills and photographic memory (for example, he could repeat the Aeneid of Virgil from beginning to end without hesitation, and for every page in the edition he could indicate which line was the first and which the last).
- Today, Euler is considered one of the greatest mathematicians of all time. His interests covered almost all aspects of Mathematics, from geometry to calculus to trigonometry to algebra to number theory, as well as optics, astronomy, cartography, mechanics, weights and measures and even the theory of music. He is also one of the most creative mathematicians and his collected works if combined can fill up to 60 to 80 quarto volumes. Leonhard Euler died of a brain hemorrhage on September 18, in 1783 at St. Petersburg, Russia. He was 76 years old. Euler was buried next to his first wife, Katharina, at the Smolensk Lutheran Cemetery.
- Euler's legacy has been enormous in terms of shaping the modern playing field of Mathematics and engineering, with his work highlighted by the Mathematical Association of America and honoured by mathematicians around the world. The asteroid 2002 Euler is also named in his honour.

Chapter 7

Euclid

The name "Euclid" means "renowned, glorious".

Born : **330 B.C**
Died : **260 B.C**
Nationality : **Swiz**

Very little information is available about Euclid but it is presumed that he was born around 330 B.C in Tyre. Going by the accounts of certain Arabic authors, he came from a rich background. His father was "Naucrates" while his grandfather was Zenarchus.

It is said that he was a Greek born in Tyre and lived in Damascus throughout his life. However, there is no certain evidence if he was the same person as Euclid of Alexandria is often confused with Euclid of Megara, another man who was a philosopher and lived at the time of Plato.

He taught in a school that he founded in Alexandria, Egypt, around 300 B.C.E. He is often referred to as the 'father of geometry' and his book Elements was used well into the 20th century as the standard textbook for teaching geometry.

Alexander of Marcedonia, the first conqueror of the world, found that the city of Alexander was the meeting place of Greeks, Jews and Arabs. Ancient Mathematics was perfected and treasured in libraries in this city. Ptolemy, the successor of Alexander, established a University and Euclid worked as teacher in this university. He taught there for about 30 years.

Contributions of Euclid to Mathematics

- His famous works are “Elements”. The elements of Euclid consist of thirteen books.
- Book I, II, IV and VI on times, areas and simple figures are mostly Pythagorean.
- Book III on circles expounds Hippocrates
- Book V on proportions and elaborates the work of Eudoxus.
- Books VII, VIII, and IX are arithmetical and give an interesting account on theory of numbers. Prime and composite numbers are introduced.
- L.C.M and G.C.D. The theory of geometrical progressions, the theorem $a^m \times a^n = a^{m+n}$ and the method of summing the progression, by use of equal ratios were also expounded by him.
- He also introduced the idea of perfect numbers, which is far harder to find than the rare postage stamps.
- A number is said to be perfect when it is equal to sum of its factors, such as 6, 23 and 486. The large perfect number so far discovered is 2^{126} (2^{127-1}).
- Book V mainly deals with the doctrine of irrational numbers, specifically of the type V $(\sqrt{a} \pm \sqrt{b})$, where a and b are positive integers.
- Book XI mainly deals with elementary solids geometry.
- Book XII deals with the method of exhaustion and finding the area of a circle in the form πr^2.
- Book XIII gives and proves the constructions for the five regular solids, and ends with decahedron.
- ‘Elements’ was Euclid’s most famous work and continues to influence Mathematics even to this day but he wrote a number of other books as well. At least 5 works of Euclid have survived to this day.
  - Data: This book holds 94 propositions and basically deals with the nature and implications of "given" information in geometrical problems.
  - On Divisions of Figures: Another important work of Euclid but survives only partially in Arabic translation. It resembles a work (3rd century) of ‘Heron of Alexandria’
  - Catoptrics: It is another important work that is related to the mathematical theory of mirrors. J J O'Connor and E F Robertson, however, consider 'Theon of Alexandria' as the real author.
  - Phaenomena: It throws light on spherical astronomy. It is strikingly similar to ‘On the Moving Sphere’ by Autolycus of Pitane, who flourished around 310 BC.

- Optics: This work shares knowledge about theory of perspective and is the earliest surviving Greek treatise on perspective.

Apart from the five above mentioned extant works, there are some other works attributed to Euclid, but have been lost. These are 'Conics', 'Porisms', 'Pseudaria' and 'Surface Loci'. In addition to these, various Arabic sources consider Euclid as author of several works on mechanics.

Expect a few logical blemishes that occur in his works, much has survived unchanged. He left nothing to be desired. Firstly he laid down carefully his definitions, then his common assumptions or axioms, then his postulates before proceeding with the orderly arrangement of their consequences. Many other writings of Euclid have come down to us dealing with astronomy, music and optics.

Euclid's Axioms

In the Elements, Euclid attempted to bring together the various geometric facts known in his day (including some that he discovered himself) in order to form an axiomatic system, in which these "facts" could be subjected to rigorous proof. His undefined terms were point, line, straight line, surface, and plane. (To Euclid, the word "line" meant any finite curve, and hence a "straight" line is what we would call a line segment.)

Euclid divided his axioms into two categories, calling the first five postulates and the next five "common notions." The distinction between postulates and common notions is that the postulates are geometric in character, whereas common notions were considered by Euclid to be true in general.

Euclid's axioms follow.

i. It is possible to draw a straight line from any point to any point.
ii. It is possible to extend a finite straight line continuously in a straight line. (In modern terminology, this says that a line segment can be extended past either of its endpoints to form an arbitrarily large line segment.)
iii. It is possible to create a circle with any center and distance (radius).
iv. All right angles are equal to one another. (A right angle is, by Euclid's definition, "half" of a straight angle: that is, if a line segment has one of its endpoints on another line segment and divides the second segment into two angles that are equal to each other, the two equal angles are called right angles.)
v. If a straight line falling on (crossing) two straight lines makes the interior angles on the same side less than two right angles, the two straight lines, if produced indefinitely, meet on that side on which the angles are less than the two right angles.
vi. Things which are equal to the same thing are equal to each other.
vii. If equals are added to equals, the wholes (sums) are equal.
viii. If equals are subtracted from equals, the remainders (differences) are equal.
ix. Things that coincide with one another are equal to one another.
x. The whole is greater than the part.

There is not much information and records relating to Euclid's personal life but historians believe that he breathed his last around 260 B.C. His most famous book "The Element" was eventually translated from Arabic to Latin by Campanus. The first printed addition of the same appeared in 1482 in Venice. In 1570, John Dee translated "The Element" in English. Dee's lectures were able to revive interest in Mathematics in England.

An Italian mathematician, Girolamo Saccheri, in 1733 tried for years to outdo the works of Euclid but his attempts to refute it went futile as he could not even find a single flaw in Euclid's theories. Eventually, he gave up and published "Euclid Cleared of Every Flaw".

The legacy that Euclid left behind is immense. He inspired personalities such as Abraham Lincoln, who used to religiously carry "The Elements" with him everywhere and used to quote the genius in his speeches.

Euclid influenced great philosophers and mathematicians such as Newton and Descartes who propounded their philosophical works using Elucid's format and structure. They also moved from simple principles towards complicated concepts just like Elucid.

Chapter 8

Pythagoras

He established and headed a society called mathematikoi.

Born : 570 BC
Died : 495 BC
Nationality : Greek

Pythagoras of Samos was a well-known mathematician, scientist and a religious teacher. He was born in Samos belonging to Johian colony and is often hailed as the first great mathematician. He settled in Crotona, a town in South Italy, 29 B.C.

Pythagoras is remembered today for his famous theorem in geometry, the 'Pythagoras Theorem'. His mentors were Thales, Pherekydes and Anaximander, who inspired him to pursue Mathematics and astronomy. Pythagoras also made important discoveries in music, astronomy and medicine. He accepted priesthood and performed the rites that were required in order to enter one of the temples in Egypt, known as Diospolis. He set up a brotherhood with some of his followers, who practiced his way of life and pursued his religious ideologies. He became one of the most distinguished teachers of religion in ancient Greece.

Pythagoras is believed to have had strong desire to learn and for this he had undertaken extensive travels. He was taught by wide range of teachers and philosophers. He spent years in Egypt in search of all available knowledge and received wisdom from an Egyptian priest Oenuphis of Heliopolis.

Contributions of Pythagoras to Mathematics

- A star shaped pentagram was the distinctive badge of brotherhood.
- In Mathematics, the Pythagoreans made great progress, particularly in the field of numbers and geometry of solids and areas.
- He noticed that vibrating strings produce harmonious tones.
- Rather Pythagoras was interested in the principles of Mathematics, the concept of number, the concept of a triangle or other mathematical figure and the abstract idea of a proof.
- Pythagoras studied properties of numbers that are familiar to mathematicians today, such as even and odd numbers, triangular numbers, perfect numbers etc.
- He discovered the theory of irrational and the construction of the cosmic figures.
- A list of theorems attributed to Pythagoras, or rather more generally to the Pythagoreans.
- The sum of the angles of a triangle is equal to two right angles. Also the Pythagoreans knew the generalization which states that a polygon with n sides has sum of interior angles 2n-4 right angles and sum of exterior angles equal to four right angles.
- The theorem of Pythagoras – for a right angled triangle, the square on the hypotenuse is equal to the sum of the squares on the other two sides.
- Constructing figures of a given area and geometrical algebra. For example they solved equation such as $a(a-x)=x^2$ by geometrical means .
- A younger mathematician Eardemus remarked "they changed Geometry into a liberal science; they diverted arithmetic from the service of commerce."
- He was interested to solve problems in application of areas.
- He introduced the term parabola, ellipse and hyperbola.
- They studied various types of progressions Arithmetic, Geometric and Harmonic.

Most of his works were lost throughout the history and very few survived. Pythagorean brotherhood couldn't survive for long as they were suppressed. According to certain historical accounts Pythagoras died along with his disciples in the temple, while some other evidences suggest that he fled to Tarentum and then to Metapontum, where he starved himself to death.

The Greek philosopher, scientist, and religious teacher Pythagoras developed a school of thought that accepted the passage of the soul into another body and established many influential mathematical and philosophical theories.

Chapter 9

Gauss

In 1810, he was honoured with the Lalande Prize by the French Academy of Sciences in recognition of his contributions to astronomv.

Born : **April 30, 1777 A.D**
Died : **February 23, 1855 A.D**
Nationality : **German**

Carl Friedrich Gauss is considered to be the greatest German mathematician of the nineteenth century. His discoveries and writings influenced and left a lasting mark in the areas of number theory, astronomy, geodesy, and Physics, particularly the study of electromagnetism.

Gauss was born in Brunswick, Germany, on April 30, 1777, to poor, working-class parents. His father laboured as a gardener and brick-layer and was regarded as an upright, honest man. However, he was a harsh parent who discouraged his young son from attending school, with expectations that he would follow one of the family trades. Luckily, Gauss' mother and uncle, Friedrich, recognized Carl's genius early on and knew that he must develop this gifted intelligence with education.

While in arithmetic class, at the age of ten, Gauss exhibited his skills as a math prodigy when the stern schoolmaster gave the following assignment: "Write down all the whole numbers from 1 to 100 and add up their sum." When each student finished, he was to bring his slate forward and place it on the schoolmaster's desk, one on top of the other. The teacher expected the beginner's class to take a good while to finish this exercise. But in a few seconds, to his teacher's surprise, Carl proceeded to the front of the room and placed his slate on the desk. Much later the other students handed in their slates.

At the end of the class time, the results were examined, with most of them wrong. But when the schoolmaster looked at Carl's slate, he was astounded to see only one number: 5,050. Carl then had to explain to his teacher that he found the result because he could see that, 1+100=101, 2+99=101, 3+98=101, so that he could find 50 pairs of numbers that each add up to 101. Thus, 50 times 101 will equal 5,050.

At the age of fourteen, Gauss was able to continue his education with the help of Carl Wilhelm Ferdinand, Duke of Brunswick. After meeting Gauss, the Duke was so impressed by the gifted student with the photographic memory that he pledged his financial support to help him continue his studies at Caroline College. At the end of his college years, Gauss made a tremendous discovery that, up to this time, mathematicians had believed was impossible. He found that a regular polygon with 17 sides could be drawn using just a compass and straight edge. Gauss was so happy about and proud of his discovery that he gave up his intention to study languages and turned to Mathematics.

Gauss' next discovery was in a totally different area of Mathematics. In 1801, astronomers had discovered what they thought was a planet, which they named Ceres. They eventually lost sight of Ceres but their observations were communicated to Gauss. He then calculated its exact position, so that it was easily rediscovered. He also worked on a new method for determining the orbits of new asteroids. Eventually these discoveries led to Gauss' appointment as professor of Mathematics and director of the observatory at Gottingen, where he remained in his official position until his death on February 23, 1855.

Carl Friedrich Gauss, though he devoted his life to Mathematics, kept his ideas, problems, and solutions in private diaries. He refused to publish theories that were not finished and perfect. Still, he is considered, along with Archimedes and Newton, to be one of the three greatest mathematicians who ever lived.

Contributions of Gauss to Mathematics

Gauss was a good Mathematician. He was born at Brunswick. He was the son of a brick-layer.

He was known of his marvellous intuition in lighter Arithmetic. His marvellous aptitude for calculation was attracted by Johan Martin Bartels, Professor of Mathematics at Darpat. Duke undertook to educate the boy and sent him to the Collegium Carolinm. When he was 19 years old, he began jotting down in a copy-book very brief Latin memoranda of his mathematical discoveries and published in 1901.

- His discovery of regular polygon of seventeen sides encouraged him to pursue Mathematics.

- Higher arithmetic was his favourite study.
- In 1807, he was appointed as the first director of a new observatory at Gotlingen.
- He wrote voluminously and made an early reputation by his work in the theory of numbers.
- He gave a result that (a log a) gives a good approximation to the total number of prime numbers less than a, when ‘a’ is large. No proof was available in his notes but Hadamard and Dela Vallee Pousin gave the proof about it fifty years ago.
- He was the last complete mathematician and he adorned every branch of science.
- He developed the theory of surface with special attention to their curvature and conditions for one surface to fit another.
- Because of the rigor imposed by Gauss, his Mathematics becomes totally different from the Mathematics of Newton, Euler and Lagrange.
- The method of “Least squares” which is indispensable at present was invented by him.
- He was the first to give a coherent account of complex numbers.
- In 1809, he published his work “Theory of motion of heavenly bodies revolving round the sun in Conic section”.
- He also made many significant advances in Geometry, Applications of Mathematics to Geodesy, Newtonian Theory of attraction etc.
- He is popularly known as “Prince of Mathematics”.

Chapter 10

Ada Lovelace

"The intellectual, the moral, the religious seem to me all naturally bound up and interlinked together in one great and harmonious whole."

Born : **December 10, 1815, London, England**
Died : **November 27, 1852, Marylebone, London**
Nationality : **British**

Ada Lovelace is the only legitimate child of the famous English poet Lord Byron. Lovelace who took an interest in hypnosis and phrenology considered herself an "analyst and metaphysician". At the age of 17, she met Charles Babbage in 1833 and collaborated with him closely. Much of her career is attributed to her outstanding contribution to Babbage's Analytical Engine.

Ada Lovelace's Personal Life

The daughter of famed poet Lord Byron, Augusta Ada Byron, Countess of Lovelace—better known as "Ada Lovelace"—was born in London on December 10, 1815. Ada showed her gift for mathematics at an early age. She translated an article on an invention by Charles Babbage, and added her own comments. Because she introduced many computer concepts, Ada is considered the first computer programmer. Ada died on November 27, 1852.

Her parents, Annabella Milbanke Byron and Lord Byron, separated briefly after her birth and although her father was allowed by the English law to get full custody of Ada, he gave up his parental rights. Annabella moved in with one-month-old Ada to Kirkby Mallory, Leicestershire.

Although her mother was the only parent she knew, the mother-daughter relationship was not very good. As a result, Ada was left in the care of her grandmother. Lord Byron died in 1824 when she was only eight years old, so she never had the chance to meet him in person. Ada Lovelace had various health problems throughout her life and in 1829 she became paralyzed after suffering from measles. She recovered gradually and in 1831, she could walk, but only with crutches.

Lovelace's Education

- Ada Lovelace had great passion for scientific developments, mathematics, and computational mathematics in particular from an early age.
- She developed the design for a flying machine when she was only 13 years old.
- She was educated by tutors, including Mary Somerville, for whom she had a great amount of respect and with whom she corresponded many years. Mary Somerville was the one who introduced her to Charles Babbage.
- Ada was encouraged by her mother to study logic and mathematics, disciplines which she believed were able to counteract the poetical tendencies that she might have inherited from her father and which Annabella considered "insanity."
- Her professor of mathematics at the University of London was Augustus de Morgan, who helped her career a great deal.

Lovelace's Contributions to Mathematics

Lovelace's work opposite British mathematician Charles Babbage, who was also a professor at Cambridge and her lifelong friend, was the most significant part of her career. Between 1842 and 1843, she translated works from French and annotated a memoir written by Italian mathematician Luigi Menabrea, who supported Babbage's Analytical Engine.

Impressed by her analytical skills and highly-developed intellectual abilities, Babbage referred to Ada as an "enchantress of numbers." In addition to being the founder of scientific computing, Lovelace aimed at creating a mathematical model that she called "a calculus of the nervous system." Her detailed annotations simply entitled Notes are the main source of her achievements throughout her life and moreover, these notes are deemed the first computer program, anticipating future developments in computer science.

In her Notes, she described an algorithm representing how the proposed mechanical general-purpose computer (the Analytical Engine) could be programmed to compute Bernoulli numbers and to solve complex problems. This is considered the first algorithm developed for implementation on a computer. In 1953, the Analytical Engine was recognized as an early model for a computer.

On July 8th, 1835, Ada married William King, 8th Baron King, and they had three children. In 1838, she became Countess of Lovelace.

Lovelace's Death

Ada Lovelace passed away on November 27, 1852, at 36 years old, due to uterine cancer complicated by excessive bloodletting. At her request, she was buried next to her father, whom she had never met, in Hucknall, Nottingham, at the Church of St. Mary Magdalene. Her proceedings on analytical engine were republished almost a hundred years after her demise.

Honour

- In 1979, a software language developed by the U.S. Department of Defense was named Ada to honour the countess' relentless services to both the disciplines of programming as well as Mathematics.
- Since 1998, the British Computer Society has awarded a medal in the countess' name and in 2008. The society initiated an annual competition for females who are students of computer science.
- Lady Lovelace has also been depicted by
    - the media in the film 'Conceiving Ada',
    - a novel 'The Difference Engine' authored by William Gibson and Bruce Sterling as well as
    - the web comic 2D Goggles designed by Sydney Padua.
- More than 150 years later, we remember her contributions to science and engineering in the celebration of Ada Lovelace Day on October 13. First celebrated in 2009 (in March), it is a day set aside to learn about women in science, technology, engineering and Mathematics.

Chapter 11

John Venn

> We endeavour to employ only symmetrical figures, such as should not only be an aid to reasoning, through the sense of sight, but should also be to some extent elegant in themselves.

Born	:	**August 4, 1834, Kingston upon Hull**
Died	:	**April 4, 1923, Cambridge**
Nationality	:	**British**

Early Life

John Venn was a British mathematician and philosopher born on 4th August 1834 in Hull, Yorkshire. His mother died when he was only three years of age. His father, who was the rector of parish of Drypool, was from a distinguished family. Venn had descended from a long line of church evangelicals. His early schooling was done from Highgate and Islington. He then went to Gonville and Caius College of Cambridge in 1853 where he got a degree in Mathematics in 1857. He became Fellow of the College; a title he kept for life.

Venn was brought up in a very strict atmosphere at home. His father Henry had played a significant part in the Evangelical movement and he was also the secretary of the 'Society for Missions to Africa and the East'. Thus he was compelled to follow the family tradition. Venn became a priest in 1859 after being ordained as a deacon at Ely. He also went to a town as a curate. However his thirst for knowledge and passion for Mathematics encouraged him to do more than this. He therefore moved back to Cambridge to lecture in moral sciences.

Work

- Venn's main area of interest was logic, and he published three texts on the subject.
- In 1866, he published The Logic of Chance, which introduced the frequency interpretation of probability; in 1881, Symbolic Logic which introduced the Venn diagrams; and in 1899, The Principles of Empirical Logic.
- Venn extended Boole's mathematical logic and is best known to mathematicians and logicians for his diagrammatic way of representing sets, and their unions and intersections. He considered three discs R, S, and T as typical subsets of a set U. The intersections of these discs and their complements divide U into eight non-overlapping regions, the unions of which give 256 different Boolean combinations of the original sets R, S, T.
- The use of geometrical representations to illustrate syllogistic logic did not originate with Venn; Gottfried Leibniz used them consistently.
- Venn wrote Symbolic Logic mostly to present his own interpretations and corrections of Boole's work.
- His fame had already been established through a paper introducing Venn diagrams on the Diagrammatic and Mechanical Representation of Prepositions and Reasonings, which appeared in the Philosophical Magazine and Journal of Science in July of 1880.
- Symbolic Logic further elaborated on these diagrams, and Venn continued to improve his method for illustrating propositions by exclusive and inclusive circles. Venn's diagrams became the most consequential part of his logic trilogy, rather than his attempt to clarify what he believed to be inconsistencies and ambiguities in Boole's logic.
- Alumni Cantabrigienses, a biographical list of all the alumni of Cambridge compiled by Venn and his son, was a massive research project which proved invaluable to historians and genealogists. The first part contained 76,000 names and covered the period up to 1751. At the time of Venn's death, the second part, covering the period from 1751 to 1900, existed in manuscript and contained a further 60,000 names.
- His aptitude for building machines led him to construct a rare machine. Its function was to bowl cricket balls. This machine turned out to be so good and accurate that once when the Australians were visiting Cambridge this machine was used on them. It actually managed to bowl out one of the top ranked player of the team four times consecutively.

Venn Diagrams

John Venn came up with Venn Diagrams in 1881. They were a representation of the relation between sets using circles within circles. For example if we take three circles A, B and C which are all subsets of D. The sections which are overlapping represent similar properties of the subsets whereas the independent areas were the individual properties of the sets. Venn diagrams can be applied in various problems however they particularly aided in the Boolean logic.

Publications

Venn has several publications to his name. The most famous of his books include

- "The Logic of Chance" (published in 1866)
- "Symbolic Logic" (published in 1881)
- "The Principles of Empirical Logic" (published in 1889).

His books on history include

- "The Biographical History of Gonville and Caius College - 1349 to 1897" (published in 1897)
- "Caius College" (published in 1901)
- "The Annals of Gonville and Caius College" (published in 1904).

Memorials

- Venn is commemorated at the University of Hull by the Venn Building, built in 1928
- A stained glass window in the dining hall of Gonville and Caius College, Cambridge, commemorates Venn's work.
- In commemoration of the 180th anniversary of Venn's birth, on 4 August 2014, Google replaced its normal logo on global search pages with an interactive and animated Google doodle that incorporated the use of a Venn diagram.
- In 1866, Venn published The Logic of Chance, a ground-breaking book which espoused the frequency theory of probability, offering that probability should be determined by how often something is forecast to occur as opposed to "educated" assumptions. Venn then further developed George Boole's theories in the 1881 work Symbolic Logic, where he highlighted what would become known as Venn diagrams.

Later Life and Death

Venn was elected as member of the Royal Society in 1883. He wrote a book 'The Biographical History of Gonville and Cauis College' which was published in 1897. He married Susanna Carnegie Edmonstone in 1867 with whom he had one son John Archibald Venn who later entered the mathematical field. He died on 4th April 1923.

Chapter 12

John Horton Conway

> People think that mathematics is complicated. Mathematics is the simple bit, it's the stuff we CAN understand. It's cats that are complicated.

Born : December 26, 1937,Liverpool, Lancashire, England
Nationality : British

Biography

Conway was born in Liverpool, the son of Cyril Horton Conway and Agnes Boyce. He became interested in mathematics at a very early age; his mother has recalled that he could recite the powers of two when he was four years old. By the age of eleven his ambition was to become a mathematician.

After leaving secondary school, Conway entered Gonville and Caius College, Cambridge to study mathematics. Conway, who was a "terribly introverted adolescent" in school, interpreted his admission to Cambridge as an opportunity to transform himself into a new person: an "extrovert".

He was awarded his Bachelor of Arts degree in 1959 and began to undertake research in number theory supervised by Harold Davenport. Having solved the open problem posed by Davenport on writing numbers as the sums of fifth powers, Conway began to become interested in infinite ordinals. It appears that his interest in games began during his years studying the Cambridge Mathematical Tripos, where he became an avid backgammon player, spending hours playing the game in the common room. He was awarded his doctorate in

1964 and was appointed as College Fellow and Lecturer in Mathematics at Cambridge University.

He left Cambridge in 1986 to take up the appointment to the John von Neumann Chair of Mathematics at Princeton University.

According to a 2015 biography, Conway has been thrice married and twice divorced, and has suffered two heart attacks, two strokes, and depression. He resides with his current wife, Diana, in Princeton.

Research

Combinatorial game theory

Among amateur mathematicians, he is perhaps most widely known for his contributions to combinatorial game theory (CGT), a theory ofpartisan games. This he developed with Elwyn Berlekamp and Richard Guy, and with them also co-authored the book Winning Ways for your Mathematical Plays. He also wrote the book On Numbers and Games (ONAG) which lays out the mathematical foundations of CGT.

He is also one of the inventors of sprouts, as well as philosopher's football. He developed detailed analyses of many other games and puzzles, such as the Soma cube, peg solitaire, and Conway's soldiers. He came up with the angel problem, which was solved in 2006.

He invented a new system of numbers, the surreal numbers, which are closely related to certain games and have been the subject of a mathematical novel by Donald Knuth. He also invented a nomenclature for exceedingly large numbers, the Conway chained arrow notation. Much of this is discussed in the 0th part of ONAG.

He is also known for the invention of Conway's Game of Life, one of the early examples of a cellular automaton. His early experiments in that field were done with pen and paper, long before personal computers existed.

Geometry

In the mid-1960s with Michael Guy, son of Richard Guy, he established that there are sixty-four convex uniform polychora excluding two infinite sets of prismatic forms. They discovered the grand antiprism in the process, the only non-Wythoffian uniform polychoron. Conway has also suggested a system of notation dedicated to describing polyhedracalled Conway polyhedron notation.

He extensively investigated lattices in higher dimensions, and determined the symmetry group of the Leech lattice.

Geometric topology

Conway's approach to computing the Alexander polynomial of knot theory involved skein relations, by a variant now called the Alexander-Conway polynomial. After lying dormant for more than a decade, this concept became central to work in the 1980s on the novel knot polynomials. Conway further developed tangle theory and invented a system of notation for tabulating knots, nowadays known as Conway notation, while correcting a number of errors in the 19th century knot tables and extending them to include all but four of the non-alternating primes with 11 crossings.

Group theory

He worked on the classification of finite simple groups and discovered the Conway groups. He was the primary author of the ATLAS of Finite Groups giving properties of manyfinite simple groups. He, along with collaborators, constructed the first concrete representations of some of the sporadic groups. More specifically, he discovered three sporadic groups based on the symmetry of the Leech lattice, which have been designated the Conway groups.

With Simon P. Norton he formulated the complex of conjectures relating the monster group with modular functions, which was named monstrous moonshine by them.

He introduced the Mathieu groupoid, an extension of the Mathieu group M12 to 13 points.

Number theory

As a graduate student, he proved the conjecture by Edward Waring that every integer could be written as the sum of 37 numbers, each raised to the fifth power, though Chen Jingrun solved the problem independently before the work could be published.

Algebra

He has also done work in algebra, particularly with quaternions. Together with Neil Sloane, he invented the icosians.

Analysis

He invented a base 13 function as a counterexample to the converse of the intermediate value theorem: the function takes on every real value in each interval on the real line, so it has a Darboux property but is not continuous.

Algorithmics

For calculating the day of the week, he invented the Doomsday algorithm. The algorithm is simple enough for anyone with basic arithmetic ability to do the calculations mentally. Conway can usually give the correct answer in under two seconds. To improve his speed, he practices his calendrical calculations on his computer, which is programmed to quiz him with random dates every time he logs on. One of his early books was on finite state machines.

Theoretical physics

In 2004, Conway and Simon B. Kochen, another Princeton mathematician, proved the free will theorem, a startling version of the 'no hidden variables' principle of quantum mechanics. It states that given certain conditions, if an experimenter can freely decide what quantities to measure in a particular experiment, then elementary particles must be free to choose their spins to make the measurements consistent with physical law. In Conway's provocative wording: "if experimenters have free will, then so do elementary particles."

Books

He has (co-)written several books including

- The ATLAS of Finite Groups
- Regular Algebra and Finite Machines
- Sphere Packings
- Lattices and Groups
- The Sensual (Quadratic) Form
- On Numbers and Games
- Winning Ways for your Mathematical Plays
- The Book of Numbers
- On Quaternions and Octonions
- The Triangle Book (written with Steve Sigur) and
- In summer 2008 published The Symmetries of Things with Chaim Goodman-Strauss and Heidi Burgiel.

Contributions to Life

Conway is the inventor of Conway's Game of Life, one of the earliest-studied and most well-known examples of a cellular automaton. He discovered many of its most fundamental and important patterns, including blinker, block, lightweight spaceship, pulsar, and R-pentomino. He was the first person to enumerate all still lifes with 7 or fewer cells.

Professional History

- 1962-1975 Lecturer in Pure Mathematics, University of Cambridge
- 1975-1983 Reader, University of Cambridge
- 1983-1987 Professor, University of Cambridge
- 1987-present von Neumann Professor, Princeton University

Visiting Positions

- 1971 Visiting Professor, California Institute of Technology
- 1984 Rademacher Lecturer, University of Pennsylvania
- 1985(Sept-Dec) University of Illinois at Chicago
- 1986-1987 Visiting von Neumann Professor, Princeton University

Honours/Awards

- 1960 Brown Prize for Pure Mathematics, Gonville and Caius College, University of Cambridge
- 1975 Junior Berwick Prize of London Mathematical Society
- 1981 Fellow of Royal Society
- 1987 Institute of Electrical and Electronic Engineering award for outstanding paper of the year
- 1987 Polya Prize of London Mathematical Society
- 1998 Frederic Esser Nemmers Prize in Mathematics

Jansi Rani J

1st Year B.Ed Student

Chapter 13

Abraham De Moivre

$$(\cos x + i \sin x)^n$$

Born : May 26, 1667
Died : November 27, 1754
Nationality : French

Abraham de Moivre was a French mathematician known for de Moivre's formula, one of those that link complex numbers and trigonometry, and for his work on the normal distribution and probability theory. He was a friend of Isaac Newton Edmond Halley, and James Stirling. Even though he faced religious persecution he remained a "steadfast Christian" throughout his life.

Early Life

Abraham De Moivre was born on 26th May 1667 in Champagne, France. His father's belief in education prompted Moivre to gain a good education but we are sure that he never gained a proper degree in Mathematics. However, we do know that he was sent to the 'Protestant Academy' in Sedan where he spent four years studying Greek. He studied logic for two years at Saumur. He also read several mathematical works such as 'Elements de Mathematiques' by Father Prestet, 'De Ratiociniis in Ludo Aleae' by Christian Huygens and a treatise on games of chances. His formal education started when he moved to Paris and took private lessons from Jacques Ozanam in 1684. Due to the beginning of religious persecution in France, De Moivre's family moved to London.

Middle Years

Abraham De Moivre started excelling in Mathematics and soon became a proficient mathematician with knowledge of the standard texts. He started tutoring Mathematics to many students at their homes and in coffee houses. He read Newton's book 'Principia' and realized he had much more to learn. He soon became profound in this text by Newton so much so that even Newton himself agreed to this by saying 'Go to Mr. De Moivre; he knows these things better than I do'.

Due to friendship with the mathematician Edmund Halleyand Isaac Newton he started gaining access to the Royal Society when his first ever paper was read to them. It was about fluxions in Principia and was published in the Philosophical Transactions that year. After the paper was published, De Moivre also generalized Newton's renowned 'Binomial Theorem' to 'Multinomial Theorem'. Impressed by this achievement, the Royal Society offered Abraham De Moivre membership which he accepted.

In the early 1700, De Moivre's attention turned to astronomy and he put forward many theories such as 'the centripetal force of any planet is directly related to its distance from the center of the forces and reciprocally related to the product of the diameter of the evolute and the cube of the perpendicular on the tangent'.

His book 'Doctrine of Chances' published in 1718 was on the subject of probability containing many innovations such as method of approximating to functions of large numbers. He is also known for his 'De Moivre's Formula' about complex numbers. During his research in the subject of probability, he also made the base for the theory of annuities. His work in this field was an expansion of the work done by his predecessors.

Later Life

Abraham De Moivre remained poor throughout his life. His income was mostly through his tutoring and some from the money he had invested in bonds. He continued his mathematical work by writing several more mathematical papers. He noticed during the last days of his life that he was sleeping an additional fifteen minutes every night. He calculated the date of his death using this information and predicted his death date to be the one when the extra fifteen minutes sleep added up to 24 hours, which turned out to be the correct date. De Moivre died on 27th November 1754 in London.

De Moivre's Formula

In 1707 de Moivre derived:

$\cos x = \frac{1}{2}(\cos(nx) + i\sin(nx))^{1/n} + \frac{1}{2}(\cos(nx) - i\sin(nx))^{1/n}$ which he was able to prove for all positive integers. In 1722 he suggested it in the more well known form of de Moivre's Formula: $(\cos x + i\sin x)^n = \cos(nx) + i\sin(nx)$

In 1749 Euler proved this formula for any real n using Euler's formula, which makes the proof quite straightforward. This formula is important because it relates complex numbers and trigonometry. Additionally, this formula allows the derivation of useful expressions for cos(nx) and sin(nx) in terms of cos(x) and sin(x).

Chapter 14

Alfred North Whitehead

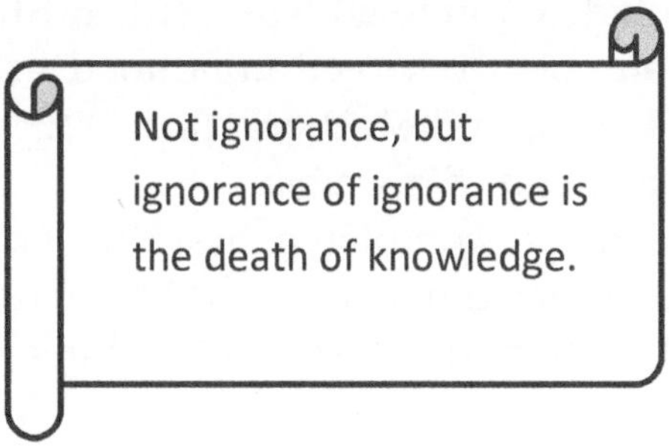

Born : **February 15, 1861, Ramsgate**
Died : **December 30, 1947, Cambridge, Massachusetts, United States**
Nationality : **British**

Alfred North Whitehead was a British mathematician who is known for his tremendous contributions in Algebra, Logic, and foundations of Mathematics, Philosophy of Science, Physics, MetaPhysics, and Education. Although Whitehead was a scientist whose areas of expertise were Maths and Physics, yet his outlook towards these subjects were philosophical than purely scientific. Whitehead was born in Britain but lived a considerable part of his life in the United States of America serving as a teacher of Philosophy at Harvard and engaged himself in a wide range of humanitarian acts.

Whitehead is also famous for co-authoring the historical "Principia Mathematica" with Bertrand Russell. Bertrand Russell and Willard Van Orman Quine had carried on with their doctoral dissertations under Whitehead's supervision. Whitehead introduced Logic and Analytic Philosophy in a brand new way. Whitehead taught Physics, the Philosophy of Science, and the theory and practice of Education at University College London and Imperial College London.

Quotes

- ❖ Not ignorance, but ignorance of ignorance is the death of knowledge.
- ❖ It requires a very unusual mind to undertake the analysis of the obvious.
- ❖ Art is the imposing of a pattern on experience, and our aesthetic enjoyment is recognition of the pattern.
- ❖ No one who achieves success does so without acknowledging the help of others. The wise and confident acknowledge this help with gratitude.
- ❖ Common sense is genius in homespun.

Childhood and Early Years

Alfred North Whitehead was born on 15 February 1869 in Ramsgate, Kent, England in an affluent family where his father and uncles were vicars. Whitehead's family was dedicated entirely to the Church of England. His grandfather was an influential man who had founded Chatham House Academy but as a kid, Whitehead was sent to Sherborne School, Dorset in South England. Whitehead got a reserved childhood at home but in school he was known to be good in sports, and Mathematics. Whitehead was bright and popular in school as he was the head of his class.

Whitehead completed his Matriculation from Trinity College, Cambridge in 1880. He received his BA in 1884. In 1884, Whitehead was chosen as the fellow of Trinity College and he started teaching and writing on Mathematics till 1910.

Career

Whitehead spent the whole of 1890s writing his mathematical book, 'Treatise on Universal Algebra' which was published in 1898. He spent the whole of 1900 with his student Bertrand Russell in coming out with the first edition of 'Principia Mathematica' which was published in 1910, 1912, and 1913. 'Principia Mathematica' was a three volume work on the foundation of Mathematics whose second edition was brought out in 1927.

Personal Life

Whitehead married Evelyn Wade, an Irish woman brought up in France, in 1890. Together the couple had three children, one daughter and two sons. He lost one of his sons while the latter served in the Royal Flying Corps during World War I. Alfred North Whitehead resigned from his position in Trinity College in 1914 in protest of the dismissal of a colleague because of an adulterous affair.

Interests

Whitehead had remained interested in theology throughout his life which resulted in his keenness to get inclined towards Roman Catholicism. It was before the World War I began that he started calling himself agnostic. Much later in his life Whitehead came back to religious beliefs.

Whitehead was pursuing his fellowship when he was deeply influenced by Physics. His dissertation examined James Clerk Maxwell's views on electricity and magnetism thus making him a Physics enthusiast.

Whitehead had been a guiding light in the field of Mathematics and Physics but his approach had remained extremely philosophical than scientific. He delved into the varied scope and nature of the subjects than into the deep clusters of tenets and theories. Such was his varied interest that he had been the President of the 'Aristotelian Society for the Systematic Study of Philosophy' from 1922 to 1923.

Contributions

Whitehead was a teacher and a notable writer of Physics, the Philosophy of Science, and the theory and practice of Education at University College London and Imperial College London from 1910 to 1926. He was a Fellow of the Royal Society since 1903 and had been elected to the British Academy in 1931.

Whitehead had formulated a rival doctrine to Einstein's general relativity. He worked on this area to come out with his more significant work, 'Enquiry Concerning the Principles of Natural Knowledge' (1919) which was an exemplary attempt to synthesize the philosophical underpinnings of Physics. Although the work has not changed the path of modern day Physics but it has been considered a great approach by Whitehead.

He addressed the Mathematical Association of England with his 'The Aims of Education' in 1916 and brought out the book with the same name in 1929. In this book and in his address Whitehead discussed and greatly criticised the existing teachers in Great Britain who neither cared for their culture nor for the students' self-education. In his popular speech Whitehead had said, "Culture is activity of thought, and receptiveness to beauty and humane feeling. Scraps of information have nothing to do with it".

In 1924 Whitehead was invited by Henry Osborn Taylor to dictate and share his ideas and also teach philosophy at Harvard University. As the subject was a soft point in Whitehead's life he rightly accepted the offer living and spending the rest of his life in the United States.

Ideas

Whitehead was a pioneer in Mathematics and Philosophy. He started viewing the world differently after the shocking collapse of Newtonian Physics which took place mostly due to Albert Einstein's work. Whitehead started out on his metaphysical views in the 1920s by bringing out 'The Concept of Nature'. He further expanded his ideas in his 1925 'Science and the Modern World' which brilliantly assesses the history of ideas and the role of Science and Mathematics in the rise of Western Civilization. In 1927 Whitehead was invited to lecture on the Gifford Lectures at the University of Edinburgh which were later published in 1929 as Process and Reality upholding process philosophy that till today remains a major contribution in Western MetaPhysics.

S.No.	Works	Year
1	A Treatise on Universal Algebra with Applications	1898
2	An Introduction to Mathematics	1911
3	The Organization of Thought Educational and Scientific	1917
4	The Principle of Relativity with Applications to Physical Science	1922
5	Principia Mathematica (with Bertrand Russell)	1925
6	Science and the Modern World	1925
7	An Enquiry Concerning the Principles of Natural Knowledge	1925
8	Religion in the Making	1926
9	Symbolism, Its Meaning and Effect	1927
10	Process and Reality: An Essay in Cosmology	1929
11	The Aims of Education and Other Essays	1929
12	Adventures of Ideas	1933
13	Nature and Life	1934
14	Modes of Thought	1938
15	Essays in Science and Philosophy	1947
16	The Wit and Wisdom of Whitehead	1947

Death

Alfred North Whitehead died on 30th December 1947 at the age of 86.

Alfred North Whitehead Timeline

- 1869 : Alfred North Whitehead was born on 15 February
- 1880 : Whitehead completed his matriculation from Trinity College, Cambridge
- 1884 : He received his BA
- 1884 : Whitehead was chosen as the fellow of Trinity College
- 1890 : Whitehead spent the whole of 1890s writing his mathematical book, 'Treatise on Universal Algebra'
- 1890 : Whitehead married Evelyn Wade, an Irish woman who was brought up in France
- 1898 : A Treatise on Universal Algebra with Applications was published
- 1903 : He had remained a Fellow of the Royal Society since
- 1910-1926 : Whitehead had been a teacher and a notable writer of Physics, the philosophy of science, and the theory and practice of education at University College London and Imperial College London
- 1911 : An Introduction to Mathematics was published
- 1917 : The Organization of Thought Educational and Scientific was published
- 1922 : He wrote 'The Principle of Relativity with Applications to Physical Science'

- ❖ 1922-1923: President of the 'Aristotelian Society for the Systematic Study of Philosophy'
- ❖ 1924: Whitehead was invited by Henry Osborn Taylor to dictate and share his ideas and also teach philosophy at Harvard University
- ❖ 1925 : He wrote Principia Mathematica with Bertrand Russell
- ❖ 1925 : Science and the Modern World was published
- ❖ 1925 : An Enquiry Concerning the Principles of Natural Knowledge was published
- ❖ 1926 : Religion in the Making was published
- ❖ 1927 : Symbolism, Its Meaning and Effect was published
- ❖ 1927: Whitehead was invited to lecture on the Gifford Lectures at the University of Edinburgh
- ❖ 1929 : Process and RealityAn Essay in Cosmology was published
- ❖ 1929 : The Aims of Education and Other Essays was published
- ❖ 1931 : He had been elected to the British Academy
- ❖ 1933 : Adventures of Ideas was published
- ❖ 1934 : Nature and Life was published
- ❖ 1938 : Modes of Thought was published
- ❖ 1947 : Essays in Science and Philosophy
- ❖ 1947 : The Wit and Wisdom of Whitehead
- ❖ 1951 : Mathematics and the Good was published
- ❖ 1954 : Dialogues of Alfred North Whitehead
- ❖ 1947 : Alfred North Whitehead died on 30th December

Chapter 15

David Hilbert

> Mathematics knows no races or geographic boundaries; for mathematics, the cultural world is one country.

Born : January 23, 1862, Konigsberg
Died : February 14, 1943, Gottingen
Nationality : German

David Hilbert was a good example of a celebrated, established, and scrupulous mathematician. His works helped pave the way for contemporary mathematical research. He was the very first to distinguish between meta-Mathematics and Mathematics. Considered one of the best mathematicians of the 20th century, Hilbert left a legacy of his vast knowledge in various divisions of Maths and was also the very first person to discover invariant theory. Today, Mathematics and Physics are still powerfully influenced by his work and his vision.

Quotes

- A mathematical theory is not to be considered complete until you have made it so clear that you can explain it to the first man whom you meet on the street.
- Mathematical Science is in my opinion an indivisible whole, an organism whose vitality is conditioned upon the connection of its parts.

- If one were to bring ten of the wisest men in the world together and ask them what the most stupid thing in existence was, they would not be able to discover anything so stupid as astrology.
- No other question has ever moved so profoundly the spirit of man; no other idea has so fruitfully stimulated his intellect; yet no other concept stands in greater need of clarification than that of the infinite.
- Mathematics knows no races or geographic boundaries; for Mathematics, the cultural world is one country.

Early Life

David Hilbert was born on January 23, 1862, in Konigsberg, Prussia, on the Baltic Sea. Prussia later merged into Germany. Since the end of World War 2, Konigsberg has been called Kaliningrad and is part of Russia.

David Hilbert's parents were Otto Hilbert, who was a judge, and Maria Therese Erdtmann. His father came from a legal family, while his mother's family was merchants. Both families were protestant, and his father held the faith rather strongly. His mother's interests shaped the young boy's interests - she was an enthusiastic amateur mathematician and astronomer.

Education

Hilbert began as a student at the Friedrichskollegium Gymnasium - a high school for academically talented children, where he studied for seven years. In his final high school year, he transferred to the more specialist math-science Wilhelm Gymnasium. He graduated from the Wilhelm Gymnasium at the highest academic level - good enough to study for a degree at any European university.

Hilbert decided to stay close to home, in 1880, he enrolled to study Mathematics at the University of Konigsberg. Five years later, he had not only obtained a degree in Mathematics, but a Ph.D. too. After completing his Ph.D. Hilbert spent winter at the University of Leipzig and then Paris. In 1886 he became a Mathematics lecturer at the University of Konigsberg.

While studying for his degrees, Hilbert made friends with two other exceptionally talented mathematicians, Hermann Minkowski, a fellow student, and Adolf Hurwitz, an Associate Professor. The three pushed one another to ever greater mathematical heights - they would continue to exchange ideas for the rest of their careers.

David Hilbert's Career

Starting in 1886, David Hilbert worked for nine years at the University of Konigsberg, first as a lecturer, then as a professor.

In 1895, he moved to become a Professor at the world's then top Mathematics University, the University of Gottingen, Germany, where giants such as Carl Friedrich Gauss, Bernhard Riemann and Peter Dirichlet had been Professors of Mathematics. Hilbert would spend the rest of his career at Gottingen.

In 1902, he became co-editor of the world's leading mathematical journal, *Mathematische Annalen.* He retired from his research and teaching works at the University of Gottingen in 1930. He continued working as co-editor of *Mathematische Annalen* until 1939.

Mathematical Achievements

Hilbert was a pure mathematician. His knowledge of Mathematics was unusually broad as well as deep, and he contributed to several areas of Mathematics and also Physics. The Mathematics he did is often at a level that can stretch the best of us, so here are brief summaries of some of his most famous achievements.

★ ***Hilbert's Basis Theorem Proof***

In 1888, Hilbert proved the finite basis theorem for any number of variables. In 1868, Paul Gordan had been able to prove the theorem, but for only two variables: three or more variables were simply too time consuming to prove. Hilbert used an entirely new abstract strategy for his proof, establishing that the theorem was true for an arbitrary number of variables. This was a major advance in algebraic number theory.

★ ***Hilbert's Axioms of Geometry***

In 1899, Hilbert published "Foundations of Geometry". Hilbert's new axioms of geometry replaced those of Euclid from over 2000 years earlier, unifying two dimensional and three dimensional geometry into one system.

Geometry, like arithmetic, requires for its logical development only a small number of simple, fundamental principles. These fundamental principles are called the axioms of geometry.

★ ***Hilbert's 23 Problems***

In 1900, Hilbert took a sweeping overview of Mathematics. He then defined his famous 23 problems. In doing so, he had a greater effect in shaping Mathematics in the 20th century than any other person. Hilbert outlined 23 problems or questions he thought, if answered correctly, would carry Mathematics to a new level. The list, he said, was not meant to exclude other problems. It was merely a sample of problems.

Sample or not, since Hilbert first posed the 23 problems, a huge amount of work has been done seeking the answers.

Mathematicians solved some of the problems within a few years, and others later, but some remain unsolved. Over one hundred years since Hilbert first listed them, the solution of these problems would still shine a bright new light on Mathematics.

The remaining great unsolved problems Hilbert identified are:

- The Riemann Hypothesis
- The Kronecker-Weber theorem extension
- The problem of the topology of algebraic curves and surfaces

John Von Neumann

"The total subject of Mathematics is clearly too broad for any of us. I do not think that any mathematician since Gauss has covered it uniformly and fully; even Hilbert did not and all of us are of considerably lesser width quite apart from the question of depth than Hilbert."

✶ *Mathematical Physics*

Hilbert was primarily a pure mathematician. Also he had broad tastes in mathematical Physics. Hilbert sometimes dabbled in this area, often as a result of discussions with Hermann Minkowski, his old friend from his student days.

After graduating, Minkowski had gone on to teach Albert Einstein in Zurich. In 1907 he had taken Einstein's Special Theory of Relativity, published in 1905, and shown that it could be advantageous to consider it differently - in four-dimensional spacetime - now called Minkowski Spacetime.

Gradually a fascination with mathematical Physics grew in Hilbert, and he spent increasing amounts of time thinking about the subject.

By 1912 it had become his primary research field. He believed that most physicists approached problems with insufficient mathematical rigour. He believed that Physics would benefit from the more rigorous approach that pure mathematicians brought to problems. He says, *"Physics is actually too hard for physicists."*

✶ *The Gravitational Field Equations of General Relativity*

In summer 1915, Albert Einstein came to Gottingen at Hilbert's invitation to lecture for a week. For years he had been struggling to express his unpublished General Theory of Relativity mathematically.

The meeting between the two great minds was obviously fruitful, because by November, they had independently derived and published field equations of gravitation, putting Einstein's General Theory of Relativity on firm mathematical ground. Hilbert and Einstein used different methods to find the equations - each method had its own strengths and weaknesses.

Hilbert never claimed any credit for the discovery of these equations, giving the credit to Einstein. Nevertheless, for some applications, Hilbert's treatment of the field equations can be rather helpful.

✶ *Hilbert Space*

Hilbert extended Vector Algebra and Calculus so they could be used in any number of dimensions. This was a huge advance in the development of both Mathematics and Physics. Today, Hilbert Space is essential in quantum mechanics, Fourier analysis, and in the application of partial differential equations, which are numerous in Physics and physical chemistry.

✶ *Logic and the Foundation of Mathematics*

In 1920 Hilbert founded mathematical formalism. He did this after realizing that there were inconsistencies at the heart of arithmetic. He hoped to repeat with arithmetic and number theory the success he had enjoyed in 1899 with the axioms of geometry.

By choosing the correct axioms, he hoped to prove that the rest of classical Mathematics would follow naturally.

In 1931, Kurt Godel was able to establish that Hilbert's Program could never be fully achieved. Godel's Incompleteness Theorems proved that there are mathematical statements which, although true, can never be mathematically proved.

Hilbert's Retirement

David Hilbert retired from his professorship in 1930. His retirement years were spent living in Nazi Germany. Jewish mathematicians, many of whom had been his friends, were banished from Gottingen: they all left for other countries.
It was a sad, rather lonely end for a brilliant, exceptionally influential mathematician, who had been a friend to everyone, and who was known for his zest for life.

When he had first arrived as a new professor at Gottingen he had upset the older professors by going to the local billiard hall, where he played against his juniors. He was worshipped by his many students, whom he made a point of going on walks with, so they could talk about mathematical problems informally.
David Hilbert died at the age of 81 on February 14, 1943, in Gottingen. Only about 10 people attended his funeral, a pitiful number for a great and much loved mathematician. For one reason or another, the Nazis had more or less cleared Gottingen's Mathematics faculty of people Hilbert had known.

After Hilbert's death, *Hermann Weyl*, a former student of Hilbert, who left Germany for America because his wife was Jewish, wrote:
David Hilbert was survived by his wife Kathe Jerosch and son Franz. Franz was intellectually challenged and suffered from one or more mental disorders. He spent some time in a mental hospital. David Hilbert found it hard to come to terms with his son's condition. Kathe Jerosch died in 1945, and Franz died in 1969.

Hermann Weyl

"No mathematician of equal stature has risen from our generation... Hilbert was singularly free from national and racial prejudices; in all public questions, be they political, social or spiritual, he stood forever on the side of freedom."
Although this ending seems rather sad, Hilbert was a man who was forever optimistic about the future of human culture and science.

Roselin Jeba M

1st Year B.Ed Student

Chapter 16

Pierre De Fermat

I have found a very great number of exceedingly beautiful theorems.

Born : **August 17, 1601 in Beaumont-de-Lomagne**
Died : **January 12, 1665, in Castres, France**
Nationality : **French**

Pierre de Fermat (17 August 1601 – 12 January 1665) was a French lawyer at the Parliament of Toulouse, France, and a mathematician who is given credit for early developments that led to infinitesimal calculus, including his technique of adequality. In particular, he is recognized for his discovery of an original method of finding the greatest and the smallest ordinates of curved lines, which is analogous to that of the differential calculus, then unknown, and his research into number theory. He made notable contributions to analytic geometry, probability, and optics. He is best known for Fermat's Last Theorem, which he described in a note at the margin of a copy of Diophantus Arithmetica.

The Life of Fermat

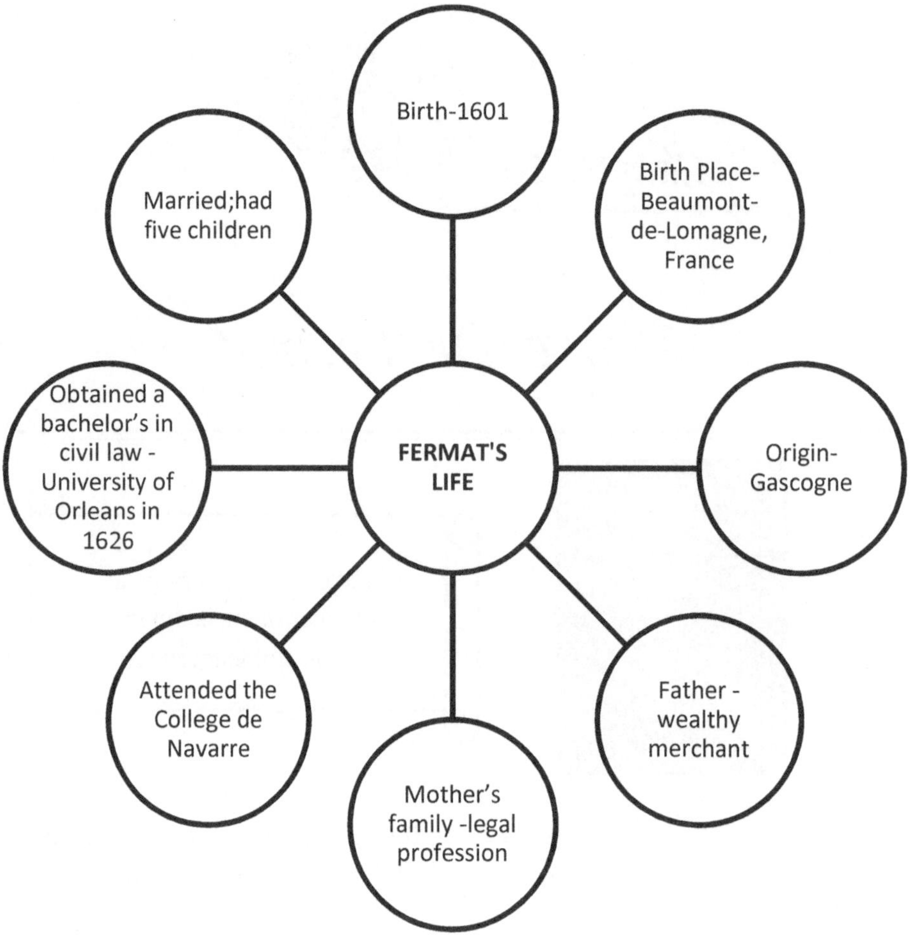

Fermat's Mathematical Research

Fermat was more of an amateur mathematician who explored the world of Mathematics as a hobby. Pierre moved to Bordeaux for post studies, where he started working on mathematical research seriously. Despite his interest in Mathematics, he always maintained it as a hobby while continuing to work as an active lawyer.

Fermat used to send his work to famous mathematicians in France. It was his connection with Marin Mersenne that gave Pierre international recognition. During his lifetime, Fermat received very marginal recognition as a mathematician and it was his papers that he shared with others that kept his work alive. Otherwise, much of his work could have been lost.

Fermat's Contributions to Mathematics

Fermat mathematician made significant contributions to number theory, probability theory, analytic geometry and the early development of infinitesimal calculus. He ventured into the areas of Mathematics which included pre-evolved calculus and trigonometry.

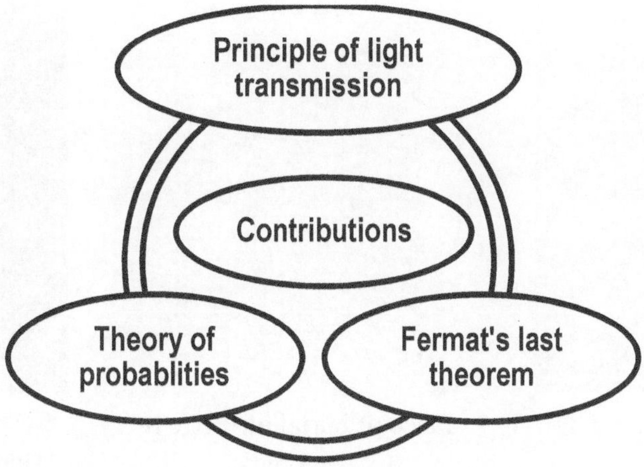

* ***Last Theorem***

Fermat's primary contribution to Mathematics was in the field of number theory. C.G. Bachet's translation of Diophantus of Alexandria inspired his interest in the Theory of Numbers. He introduced Fermat's "Last Theorem," which states that *there is no solution in integers of the equation xn + yn = zn (xyz# 0, n>2).*

Fermat contributed to the development of calculus through his work on the properties of curves. Sir Isaac Newton said that *his invention of calculus was based on Fermat's methods of tangents.* Fermat's work on calculus was an aid in developing the differential calculus.

* ***Theory of Probabilities***

Fermat, along with Pascal, is known as the *founder of Theory of Probabilities.* His views on fundamental principles of the subject became the foundation of the probability theory. Fermat's theory of probabilities grew out of his early research into the theory of numbers.

* ***Principle of Light Transmission***

Fermat had a strong interest in maximum-minimum problems, which he applied in the field of optics. He applied his method of maxima and minima and made the assumption that the speed of the light is more in the denser medium involved in refraction. From this principle, Fermat deduced the laws of refraction and reflection. This principle was known as Fermat's Principle.

Most of the history's thinkers are remembered for their completed work. Pierre did not publish during his lifetime regarding his findings. By any standards, Pierre Fermat was a great mathematician, but he is best remembered not for what he did, but for what he left undone.

Statue of Fermat

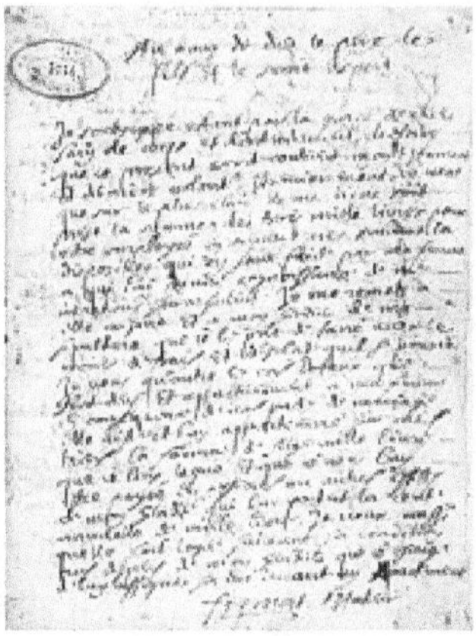

Place of burial of Pierre de Fermat

Holographic Will Handwritten by Fermat

Translation of the plaque: in this place was buried on January 13, 1665, Pierre de Fermat, councilor of the chamber and mathematician of great renown, celebrated for his theorem, $an + bn \neq cn$ *for* $n>2$

Holographic Will handwritten by Fermat on 4 March 1660 is kept at the Departmental Archives of Haute-Garonne, in Toulouse.

Assessment of his work

Together with Rene Descartes, Fermat was one of the two leading mathematicians of the first half of the 17th century. According to Peter L. Bernstein, in his book 'Against the Gods', Fermat was a mathematician of rare power. He was an independent inventor of analytic geometry; he contributed to the early development of calculus; he did research on the weight of the earth and he worked on light refraction and optics. In the course of what turned out to be an extended correspondence with Pascal, he made a significant contribution to the theory of probability. But Fermat's crowning achievement was in the theory of numbers."

Things named after Fermat

- Fermat-Apollonius circle
- Fermat-Catalan conjecture
- Fermat cubic
- Fermat curve
- Fermat number
- Fermat point
- Fermat polygonal number theorem
- Fermat primality test
- Fermat Prize
- Fermat pseudoprime
- Fermat quintic threefold
- Fermat quotient
- Fermat's factorization method
- Fermat's last theorem
- Fermat's little theorem
- Fermat's principle
- Fermat's spiral
- Fermat's theorem (stationary points)
- Fermat's theorem on sums of two squares

Chapter 17

Fibonacci

Fibonacci numbers were used in Indian Mathematics 1500 years before Fibonacci was even born.

Born : c.1170, Pisa
Died : c.1250, Pisa
Nationality : Italian

Leonardo of Pisa (Fibonacci) (1170-1250)

The 13th Century Italian Leonardo of Pisa, better known by his nickname *Fibonacci*, was the most talented Western mathematician of the middle ages. He was the son of a customs official and as a child he travelled around North Africa with his father, where he learned about Arabic Mathematics.

In particular, in 1202, he wrote a hugely influential book called **"Liber Abaci" ("Book of Calculation"),** in which he promoted the use of the Hindu-Arabic numeral system. Despite its obvious advantages, uptake of the system in Europe was slow, and Arabic numerals were even banned in the city of Florence in 1299 on the pretext that they were easier to falsify than Roman numerals. However, common sense eventually prevailed and the new system was adopted throughout Europe by the 15th century, making the Roman system obsolete. The horizontal bar notation for fractions was also first used in this work.

His Books

Liber Abbaci (The Book of Calculation), 1202 (1228)
Practica Geometriae (The Practice of Geometry), 1220
Liber Quadratorum (The Book of Square Numbers), 1225
Flos (The Flower), 1225
Letter to Master Theodore

History of Fibonacci sequence

Fibonacci is known for his introduction of a particular number sequence Fibonacci Numbers or the Fibonacci Sequence. He discovered the sequence - the first recursive number sequence known in Europe - while considering a practical problem in the "Liber Abaci" involving the growth of a hypothetical population of rabbits based on idealized assumptions. He noted that, after each monthly generation, the number of pairs of rabbits increased from 1 to 2 to 3 to 5 to 8 to 13, etc, and identified how the sequence progressed by adding the previous two terms (in mathematical terms, $F_n = F_{n-1} + F_{n-2}$, a sequence which could in theory extend indefinitely.

Fact behind Fibonacci numbers

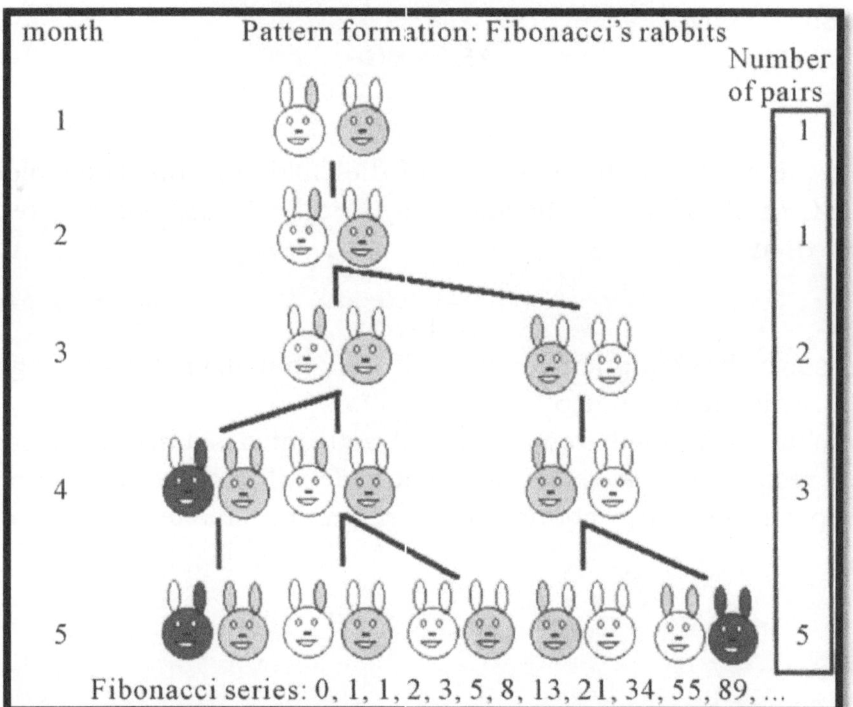

The Golden Ratio

The golden ratio is a special number approximately equal to 1.6180339887498948482. We use the Greek letter Phi to refer to this ratio. Like Pi, the digits of the Golden Ratio go on forever without repeating. It is often better to use its exact value:

$$\frac{1 + \sqrt{5}}{2}$$

The Golden Rectangle

A Golden Rectangle is a rectangle in which the ratio of the length to the width is the Golden Ratio. In other words, if one side of a Golden Rectangle is 2 ft. long, the other side will be approximately equal to 2 * (1.62) = 3.24.

Now that you know a little about the Golden Ratio and the Golden Rectangle, let's look a little deeper. Take a line segment and label its two endpoints A and C. Now put a point B between A and C so that the ratio of the short part of the segment (AB) to the long part (BC) equals the ratio of the long part (BC) to the entire segment (AC):

The ratio of the lengths of the two parts of this segment is the Golden Ratio. In an equation, we have

$$\frac{AB}{BC} = \frac{BC}{AC}.$$

Now we're ready for the definition of the Golden Ratio. The Golden Ratio is the ratio of BC to AB. If we set the value of AB to be 1, and use x to represent the length of BC, then

$$\frac{1}{x} = \frac{x}{1+x}.$$

If we solve this equation for x, we'll find that it is the value given above, (1+sqrt{5})/2, which is about 1.62.

If you have a Golden Rectangle and you cut a square off it so that what remains is a rectangle, that remaining rectangle will also be a Golden Rectangle. You can keep cutting these squares off and getting smaller and smaller Golden Rectangles.

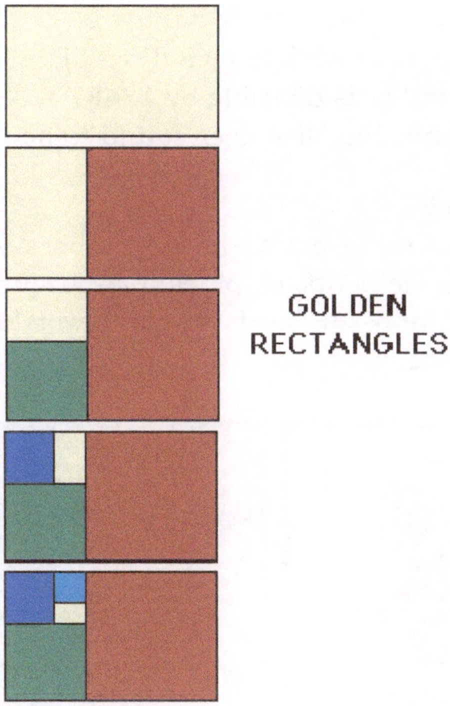

Fibonacci Sequence

In the Fibonacci Sequence (0, 1, 1, 2, 3, 5, 8, 13, ...), each term is the sum of the two previous terms (for instance, 2+3=5, 3+5=8, ...). As you go farther and farther to the right in this sequence, the ratio of a term to the one before it will get closer and closer to the Golden Ratio.

With the Fibonacci Sequence you can do the opposite of what we described above for the Golden Rectangle. Start with a square and add a square of the same size to form a new rectangle. Continue adding squares whose sides are the length of the longer side of the rectangle; the longer side will always be a successive Fibonacci number. Eventually the large rectangle formed will look like a Golden Rectangle - the longer you continue, the closer it will be.

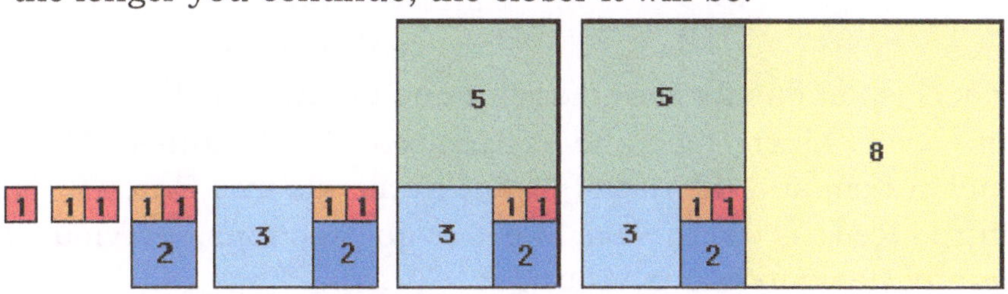

FIBONACCI SQUARES

Properties of Fibonacci sequence

Many of the implications and relationships of the sequence were not discovered until several centuries after Fibonacci death.

For instance, the sequence regenerates itself in some surprising ways:

- φ Every third F-number is divisible by 2 (F3 = 2),
- φ Every fourth F-number is divisible by 3 (F4 = 3),

- φ Every fifth F-number is divisible by 5 (F5 = 5),
- φ Every sixth F-number is divisible by 8 (F6 = 8),
- φ Every seventh F-number is divisible by 13 (F7 = 13), etc.

The numbers of the sequence has also been found to be ubiquitous in nature.

Applications of Fibonacci

Many species of flowering plants have numbers of petals in the Fibonacci sequence, the spiral arrangements of pineapples occur in 5s and 8s, those of pinecones in 8s and 13s, and the seeds of sunflower heads in 21s, 34s, 55s or even higher terms in the sequence; etc.

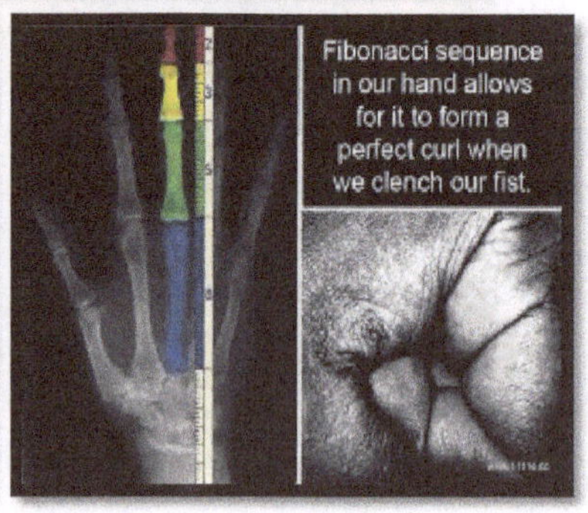

Contributions

Fibonacci is famous for his contributions to number theory.

- In his book, Liber abaci he introduced the Hindu-Arabic place-valued decimal system and the use of Arabic numerals into Europe.
- He introduced us to the bar we use in fractions, previous to this, the numerator has quotations around it.
- The square root notation is also a Fibonacci method.

Chapter 18

Omarkhayyam

Be happy for this moment. This moment is your life

Born : **May 18, 1048, Nishapur, Iran**
Died : **December 4, 1131, Nishapur, Iran**
Nationality : **Persian**

Early life

Omarkhayyam was born in Nishapur, in Iran, then a Seljuq capital in Khorasan, which rivaled Cairo or Baghdad in cultural prominence in that era. He is thought to have been born into a family of tent-makers (khayy m "tent-maker"), which he would make into a play on words later in life.

Omar Khayyam

He spent part of his childhood in the town of Balkh, studying under the well-known scholar Sheikh Muhammad Mansuri. He later studied under one of the greatest teachers of the Khorasan region Imam Mowaffaq Nishapuri.

Omar Khayyam's years in Isfahan were very productive ones, but after the death of the Seljuq Sultan Malik-Shah I, the Sultan's widow turned against him as an adviser, and as a result, he soon set out on his Hajj or pilgrimage to Mecca and Medina. He was then allowed to work as a court astrologer, and was permitted to return to Nishapur, where he was renowned for his works, and continued to teach Mathematics, astronomy and even medicine.

Contributions to Mathematics

- Khayyam wrote the influential Treatise on Demonstration of Problems of Algebra (1070), which laid down the principles of algebra, part of the body of Mathematics that was eventually transmitted to Europe.
- He derived general methods for solving cubic equations and even some higher orders.
- In the Treatise, he wrote on the triangular array of binomial coefficients known as Pascal's triangle.
- In 1077, Khayyam wrote **Sharh ma ashkala min musadarat kitab Uqlidis** (Explanations of the Difficulties in the Postulates of Euclid) published in English as "On the Difficulties of Euclid's Definitions".
- He has contributed to the eventual development of non-Euclidean geometry.
- Omar Khayyam created important works on geometry, specifically on the theory of proportions.

Omar Khayyam Treatise on Demonstration of Problems of Algebra

This particular remark of Khayyam and certain propositions found in his Algebra book have made some historians of Mathematics believe that Khayyam had indeed a binomial theorem up to any power. The case of power 2 is explicitly stated in Euclid's elements and the case of at most power 3 had been established by Indian mathematicians. Khayyam was the mathematician who noticed the importance of a general binomial theorem. The argument supporting the claim that Khayyam had a general binomial theorem is based on his ability to extract roots.

Khayyam-Saccheri quadrilateral

The Saccheri quadrilateral was first considered by Khayyam in the late 11th century in Book I of Explanations of the Difficulties in the Postulates of Euclid. Khayyam was not trying to prove the parallel postulate as such but to derive it from an equivalent postulate he formulated from "the principles of the Philosopher" (Aristotle):

Two convergent straight lines intersect and it is impossible for two convergent straight lines to diverge in the direction in which they converge.

Astronomer

The ***Jalali calendar*** was introduced by Omar Khayyam. Today it is one of the oldest calendars in the world as well as the most accurate solar calendar in use today. Since the calendar uses astronomical calculation for determining the vernal equinox, it has no intrinsic error, but this makes it an observation based calendar.

Like most Persian mathematicians of the period, Khayyam was also an astronomer and achieved fame in that role. His calendar was more accurate to the mean tropical year than the Gregorian calendar of 500 years later. The modern Iranian calendar is based on his calculations.

Heliocentric theory

It is sometimes claimed that Khayyam demonstrated that the earth rotates on its axis by presenting a model of the stars to a planetarium. The other source for the claim that Khayyam believed in heliocentrism is Edward FitzGerald's popular but anachronistic rendering of Khayyam's poetry, in which the first lines are mistranslated with a heliocentric image of the Sun flinging "the Stone that puts the Stars to Flight".

Legacy

- A lunar crater Omar Khayyam was named after him in 1970
- A minor planet called 3095 Omarkhayyam, discovered by Soviet astronomer Lyudmila Zhuravlyova in 1980, is named after him.
- In June 2009 Iran donated a scholar pavilion to United Nations Office in Vienna which is placed in the central Memorial Plaza of the Vienna International Center.

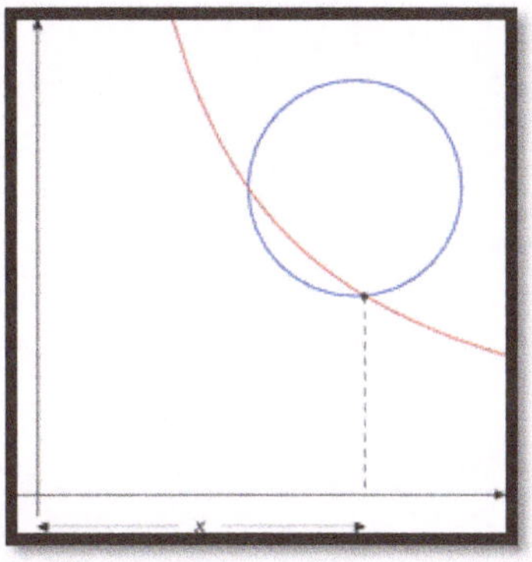

Omar Khayyam's geometric solution to the cubic equation $x^3 + 200x = 20x^2 + 2000$.

Statue of Omar Khayyam in Bucharest

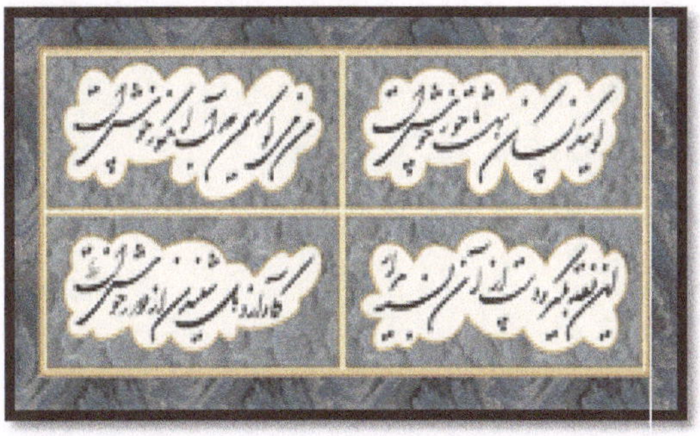

A well decorated plaque containing poems from the Rubáiyát of Omar Khayyam

An Ottoman Era inscription of a poem written by Omar Khayyam at Morica Han in Sarajevo, Bosnia and Herzegovina

Vinitha Lesslin Agila D

1st Year B.Ed Student

Chapter 19

Dattatreya Ramachandra Kaprekar

Kaprekar is known for his Recreational Maths

Born : **January 17, 1905, Dahanu**
Died : **1986 in Devlali, India**
Nationality : **Indian**

Dattaraya Ramchandra Kaprekar (1905–1986) was an Indian recreational mathematician who described several classes of natural numbers including the Kaprekar, Harshad and Self Numbers and discovered the Kaprekar Constant, named after him.

Despite having no formal postgraduate training and working as a school teacher, he published extensively and became well known in recreational Mathematics circles. Kaprekar received his secondary school education in Thane and studied at Fergusson College in Pune. In 1927 he won the Wrangler R. P. Paranjpe Mathematical Prize for an original piece of work in Mathematics. He attended the University of Mumbai, receiving his bachelor's degree in 1929. Having never received any formal postgraduate training, for his entire career (1930–1962) he was a school teacher at Nasik in Maharashtra, India. He published extensively on topics like recurring decimals, magic squares, and integers with special properties. He is also known as "Ganitanand".

Contributions for Mathematics

Today his name is well-known and many other mathematicians have pursued the study of the properties he discovered.

Kaprekar constant

- In **1949**, Kaprekar discovered an interesting property of the number **6174**, which was subsequently named the Kaprekar constant.
- He showed that 6174 is reached in the limit as one repeatedly subtracts the highest and lowest numbers that can be constructed from a set of four digits that are not all identical.
- Thus, starting with 1234, we have
  - 4321 - 1234 = 3087, then
  - 8730 - 0378 = 8352, and
  - 8532 - 2358 = 6174.

Repeating from this point onward leaves the same number (7641 - 1467 = 6174). In general, when the operation converges it does so in at most seven iterations. A similar constant for 3 digits is 495. However, in base 10 a single such constant only exists for numbers of 3 or 4 digits; for more digits (or 2), the numbers enter into one of several cycles.

Kaprekar number

- A Kaprekar number is a positive integer with the property that if it is squared, then its representation can be partitioned into two positive integer parts whose sum is equal to the original number (e.g. 45, since 45^2=2025, and 20+25=45, also 9, 55, 99 etc.) However, note the restriction that the two numbers are positive; for example, 100 is not a Kaprekar number even though 100^2=10000, and 100+00 = 100.
- This operation, of taking the rightmost digits of a square, and adding it to the integer formed by the leftmost digits, is known as the **Kaprekar operation.**

Some examples of Kaprekar numbers in base 10, besides the numbers 9, 99, 999... are (sequence A006886 in OEIS):

Number	Square	Decomposition
703	703^2 = 494209	494+209 = 703
2728	2728^2 = 7441984	744+1984 = 2728
5292	5292^2 = 28005264	28+005264 = 5292
857143	857143^2 = 734694122449	734694+122449 = 857143

Devlali or Self number

- In 1963, Kaprekar defined the property which has come to be known as self-numbers, which are integers that cannot be generated by taking some other number and adding its own digits to it.
- For example, 21 is not a self-number, since it can be generated from 15: 15 + 1 + 5 = 21.
- But 20 is a self-number, since it cannot be generated from any other integer. He also gave a test for verifying this property in any number.
- These are sometimes referred to as Devlali numbers (after the town where he lived); though this appears to have been his preferred designation, the term self-number is more widespread.
- Sometimes these are also designated Colombian numbers after a later designation.

Harshad number

- Kaprekar also described the Harshad numbers which he named harshad, meaning "**giving joy**"; these are defined by the property that they are divisible by the sum of their digits.
- Thus 12, which is divisible by 1 + 2 = 3, is a Harshad number. These were later also called Niven numbers after a 1977 lecture on these by the Canadian mathematician Ivan M. Niven.
- Numbers which are Harshad in all bases (only 1, 2, 4, and 6) are called **Harshad numbers**.
- Much work has been done on Harshad numbers, and their distribution, frequency, etc. are a matter of considerable interest in number theory today.

Demlo number

- Kaprekar also studied the Demlo numbers, named after a train station 30 miles from Bombay on the then G. I. P. Railway where he had the idea of studying them. These are the numbers 1, 121, 12321, ..., which are the squares of the repunits 1, 11, 111, ...

A drunkard wants to go on drinking wine to remain in that pleasurable state. The same is the case with me in so far as numbers are concerned.

Kaprekar invented different number properties throughout his life. He was not well known, however, despite many of his papers being reviewed in Mathematical Reviews. International fame only came in 1975 when Martin Gardener wrote about Kaprekar and his numbers in his 'Mathematical Games' column in the March issue of Scientific American.

Chapter 20

Prasanta Chandra Mahalanobis

His birthday, June 29 is celebrated as National Statistical Day.

Born : **June 29, 1893**
Died : **June 28, 1972**
Nationality : **Indian**

- **Prasanta Chandra Mahalanobis** was an Indian scientist and applied statistician.
- He is best remembered for the Mahalanobis distance, a statistical measure.
- He made pioneering studies in anthropometry in India.
- He founded the Indian Statistical Institute, and contributed to the design of large-scale sample survey.

Early Life

Mahalanobis belonged to a family of Bengali landed gentry who lived in Bikrampur (now in Bangladesh). His grandfather Gurucharan (1833–1916) moved to Calcutta in 1854 and built up a business, starting a chemist shop in 1860. Gurucharan was influenced by Debendranath Tagore (1817–1905), father of the Nobel Prize–winning poet, Rabindranath Tagore.

His house on 210 Cornwallis Street was the center of the Brahmo Samaj. Gurucharan married a widow against social traditions. His elder son

Subodhchandra (1867–1953) was a distinguished educationist who studied physiology at Edinburgh University. He was a Fellow of the Royal Society of Edinburgh. He was the Head of the Dept. of Physiology, University of Cardiff (first Indian to occupy this post in a British university).

In 1900, he founded Dept. of Physiology in the Presidency College, Calcutta. Subodhchandra also became a member of the Senate of the Calcutta University. Gurucharan's younger son, **Prabodh Chandra** (1869-1942) was the father of P. C. Mahalanobis.

Born in the house at 210 Cornwallis Street, P. C. Mahalanobis, grew up in a socially active family surrounded by intellectuals and reformers. Mahalanobis received his early schooling at the Brahmo Boys School in Calcutta graduating in 1908. He then joined the Presidency College, Calcutta where he was taught by teachers who included Jagadish Chandra Bose, and Prafulla Chandra Ray. Subhas Chandra Bose was two years his junior at college. Mahalanobis received a Bachelor of Science degree with honours in Physics in 1912.

He interacted with the mathematical genius **Srinivasa Ramanujan** during the latter's time at Cambridge. After his Tripos in Physics, Mahalanobis worked with C. T. R. Wilson at the Cavendish Laboratory. He discovered the utility of statistics to problems in meteorology, anthropology and began working on it on his journey back to India in Calcutta.

Indian Statistical Institute (ISI)

- Many colleagues of Mahalanobis took an interest in statistics and the group grew in the Statistical Laboratory located in his room at the Presidency College, Calcutta.
- A meeting was called on 17th December 1931 with Pramatha Nath Banerji (Minto Professor of Economics), Nikhil Ranjan Sen (Khaira Professor of Applied Mathematics) and Sir R. N. Mukherji.
- The meeting led to the establishment of the ISI, and formally registered on 28th April 1932 as a non-profit distributing learned society under the Societies Registration Act XXI of 1860.
- ISI was initially in the Physics Department of the Presidency College and the expenditure in the first year was Rs. 238.
- It gradually grew with the pioneering work of a group of his colleagues including S. S. Bose, J. M. Sengupta, R. C. Bose, S. N. Roy, K. R. Nair, R.R. Bahadur, Gopinath Kallianpur, D.B.Lahir and C.R.Rao.

Contribution to Statistics

Mahalanobis distance

- **Mahalanobis distance** A chance meeting with Nelson Annandale, then the director of the Zoological Survey of India, at the 1920 Nagpur session of the Indian Science Congress led to Annandale asking him to analyse anthropometric measurements of Anglo-Indians in Calcutta.

- Mahalanobis had been influenced by the anthropometric studies published in the journal *Biometrical* and he chose to ask the questions on what factors influence the formation of European and Indian marriages.
- He wanted to examine if the Indian side came from any specific castes. He used the data collected by Annandale and the caste specific measurements made by Herbert Risley to come up with the conclusion that the sample represented a mix of Europeans mainly with people from Bengal and Punjab but not with those from the Northwest Frontier Provinces or from Chhota Nagpur.
- He also concluded that the intermixture more frequently involved the higher castes than the lower ones. This analysis was described by his first scientific paper in 1922.
- During the course of these studies he found a way of comparing and grouping populations using a multivariate distance measure. This measure, denoted "D2" and now eponymously named Mahalanobis distance, is independent of measurement scale.
- Biologist and statistician Ronald Fisher inspired by biometrical and mentored by Acharya Brajendra Nath Seal started his statistical work. Initially he worked on analysing university exam results, anthropometric measurements on Anglo-Indians of Calcutta and some meteorological problems.
- He also worked as a meteorologist for some time. In 1924, when he was working on the probable error of results of agricultural experiments, he met Ronald Fisher and Anil Kumar Gain, with whom he established a lifelong friendship. He also worked on schemes to prevent floods.

Sample surveys

- His most important contributions are related to large scale sample surveys. He introduced the concept of pilot surveys and advocated the usefulness of sampling methods.
- Early surveys began between 1937 and 1944 and included topics such as consumer expenditure, tea-drinking habits, public opinion, crop acreage and plant disease.
- *Harold Hotelling wrote: "No technique of random sample has, so far as I can find, been developed in the United States or elsewhere, which can compare in accuracy with that described by Professor Mahalanobis" and Sir R. A. Fisher commented that "The ISI has taken the lead in the original development of the technique of sample surveys, the most potent fact finding process available to the administration".*
- He introduced a method for estimating crop yields which involved statisticians sampling in the fields by cutting crops in a circle of diameter 4 feet.

- Others such as P. V. Sukhatme and V. G. Panse who began to work on crop surveys with the Indian Council of Agricultural Research and the Indian Agricultural Statistics Research Institute suggested that a survey system should make use of the existing administrative framework. The differences in opinion led to acrimony and there was little interaction between Mahalanobis and agricultural research in later years.

Linguistics

- Mahalanobis also started research in the field of quantitative linguistics and language planning in the Linguistic Research Unit of the Indian Statistical Institute.
- He also worked on Speech Pathology in collaboration with Djordge Kostic, Rhea Das and Alakananda Mitter and made some contributions to the field of language correction.

Later life

- In later life, Mahalanobis was a member of the Planning Commission. He contributed prominently to newly independent India's five-year plans starting from the second.
- In the second five-year plan he emphasised industrialisation on the basis of a two-sector model. His variant of Wassily Leontief's Input-Output model, the Mahalanobis model, was employed in the Second Five Year Plan, which worked towards the rapid industrialisation of India and with other colleagues at his institute; he played a key role in the development of a statistical infrastructure.
- He encouraged a project to assess deindustrialisation in India and correct some previous census methodology errors and entrusted this project to Daniel Thorner.
- He also had an abiding interest in cultural pursuits and served as secretary to Rabindranath Tagore, particularly during the latter's foreign travels, and also worked at his Visva-Bharati University, for some time.
- He received one of the highest civilian awards, the **Padma Vibhushan** from the Government of India for his contribution to science and services to the country.

Honours

S.No.	Honours	Year
1	Weldon Memorial Prize from the University of Oxford	1944
2	Fellow of the Royal Society, London	1945
3	President of Indian Science Congress	1950
4	Fellow of the Econometric Society, USA	1951
5	Fellow of the Pakistan Statistical Association	1952
6	Honourary Fellow of the Royal Statistical Society, UK	1954
7	Sir Deviprasad Sarvadhikari Gold Medal	1957
8	Foreign member of the Academy of Sciences of the USSR	1958
9	Honourary Fellow of King's College, Cambridge	1959
10	Fellow of the American Statistical Association	1961
11	Durgaprasad Khaitan Gold Medal	1961
12	Padma Vibhushan	1968
13	Srinivasa Ramanujan Gold Medal	1968

Mahalanobis died on 28th June 1972, a day before his seventy-ninth birthday. Even at this age, he was still active doing research work and discharging his duties as the Secretary and Director of the Indian Statistical Institute and as the Honourary Statistical Advisor to the Cabinet of the Government of India.

The government of India decided in 2006 to celebrate his birthday, 29 June, as *National Statistical Day.*

Chapter 21

Satyendranath Bose

He is best known for his work on quantum mechanics

Born : January 1, 1894
Died : February 4,1974
Nationality : Indian

Satyendra Nath Bose was a noted Indian mathematician, physicist whose contribution holds special in the field of Indian Science. He is best known for his work on quantum mechanics in the early 1920s, providing the foundation for Bose–Einstein statistics and the theory of the Bose–Einstein condensate.

A Fellow of the Royal Society, he was awarded India's second highest civilian award, the Padma Vibhushan in 1954 by the Government of India. The class of particles that obey Bose–Einstein statistics, bosons, was named after Bose by Paul Dirac. A self-taught scholar and a polyglot, he had a wide range of interests in varied fields including Physics, Mathematics, chemistry, biology, mineralogy, philosophy, arts, literature, and music. He served on many research and development committees in independent India.

Early life

Bose was born in Calcutta (now Kolkata), the eldest of seven children. He was the only son, with six sisters after him. His ancestral home was in village

Bara Jagulia, in the district of Nadia, in the state of West Bengal. His schooling began at the age of five, near his home.

When his family moved to Goabagan, he was admitted to the New Indian School. In the final year of school, he was admitted to the Hindu School. He passed his entrance examination (matriculation) in 1909 and stood fifth in the order of merit. He next joined the intermediate science course at the Presidency College, Calcutta, where he was taught by illustrious teachers such as Jagadish Chandra Bose, Sarada Prasanna Das, and Prafulla Chandra Ray.

Naman Sharma and Meghnad Saha, from Dacca (Dhaka), joined the same college two years later. Prasanta Chandra Mahalanobis and Sisir Kumar Mitra were few years senior to Bose. Satyendra Nath Bose chose mixed (applied) Mathematics for his B.Sc and passed the examinations standing first in 1913 and again stood first in the M.Sc mixed Mathematics exam in 1915. It is said that his marks in the M.Sc examination created a new record in the annals of the University of Calcutta, which is yet to be surpassed.

After completing his M.Sc, Bose joined the University of Calcutta as a research scholar in 1916 and started his studies in the theory of relativity. Quantum theory had just appeared on the horizon and important results had started pouring in. His father, Surendranath Bose, worked in the Engineering Department of the East Indian Railway Company. In 1914, age 20, Satyendra Nath Bose married Ushabati Ghosh, the 11-year-old daughter of a prominent Calcutta physician. They had nine children, but two died in early childhood. When he died in 1974, he left behind his wife, two sons, and five daughters.

As a polyglot, he was well versed in several languages such as Bengali, English, French, German and Sanskrit as well as the poetry of Lord Tennyson, Rabindranath Tagore and Kalidasa. He could also play a musical instrument similar to a violin. He was actively involved in running night schools that came to be known as the Working Men's Institute.

Research career

- Bose attended Hindu School in Calcutta, and later attended Presidency College, also in Calcutta, earning the highest marks at each institution, while fellow student and future astrophysicist Meghnad Saha came second.
- He came in contact with teachers such as Jagadish Chandra Bose, Prafulla Chandra Ray and Naman Sharma who provided inspiration to aim high in life.
- From 1916 to 1921, he was a lecturer in the Physics department of the University of Calcutta.
- Along with Saha, Bose prepared the first book in English based on German and French translations of original papers on Einstein's special and general relativity in 1919.
- In 1921, he joined as Reader of the department of Physics of the recently founded University of Dhaka (in present-day Bangladesh).

- Bose set up whole new departments, including laboratories, to teach advanced courses for M.Sc and B.Sc honours and taught thermodynamics as well as James Clerk Maxwell's theory of electromagnetism.
- Satyendra Nath Bose, along with Saha, presented several papers in theoretical Physics and pure Mathematics from 1918 onwards.
- In 1924, while working as a Reader (Professor without a chair) at the Physics Department of the University of Dhaka, Bose wrote a paper deriving Planck's quantum radiation law without any reference to classical Physics by using a novel way of counting states with identical particles.
- This paper was seminal in creating the very important field of quantum statistics. Though not accepted at once for publication, he sent the article directly to Albert Einstein in Germany.

S .N .Bose with other scientists at Calcutta University

Major Works

- Satyendra Nath Bose is best known for giving the concepts of 'Boson', which refers to one of the two classes of particles. His work in quantum Physics was further developed by Albert Einstein which laid the foundation for Bose-Einstein statistics and the theory of the Bose-Einstein condensate.

Awards and Achievements

- The Government of India bestowed upon this eminent physicist the title of Padma Vibhushan in 1954 for his services towards science and research.
- The S.N. Bose National Centre for Basic Sciences was established by the government in Calcutta in 1986.

Personal Life and Legacy

- He got married to Ushabati when he was 20 years old. The couple had nine children of whom two died in their childhood.
- He died at the age of 80 in 1974. He was survived by his wife of 60 years and seven children at the time of his death.

Sakthi Meena K

1st Year B.Ed Student

Chapter 22

Daina Taimina

Thank you for your Crocheting Adventures

Born : August 19, 1954, Riga
Nationality : Latvian Mathematics

Biography

She received all her formal education in Riga, Latvia, where in 1977 she graduated summa cum laude from the University of Latvia and completed her graduate work in theoretical Computer Science in 1990. At that time, a doctoral thesis had to be defended outside of Latvia, so she defended hers in Minsk. This explains the fact that formally Taimina's doctorate was issued by the Institute of Mathematics of the National Academy of Sciences of Belarus. After Latvia regained independence in 1991, Taimina received her doctorate in Mathematics from the University of Latvia, where she taught for 20 years.

Daina Taimina joined the Cornell Math Department in December 1996. While attending a geometry workshop in 1997, she saw fragile paper models of hyperbolic planes, designed by geometer William Thurston. She decided to make more durable models, and did so by crocheting them. Due to her success in this she was invited, together with her husband David Henderson, a math professor also at Cornell, to give a presentation at a Cornell workshop. Crocheted mathematical models later appeared in three geometry textbooks they wrote together, of which the most popular is Experiencing Geometry: Euclidean and non-Euclidean with History.

Written Books

An article about Taimina's innovation in New Scientist was spotted by the Institute For Figuring, a small non-profit organisation based in Los Angeles, and she was invited to speak about hyperbolic space and its connections with nature to a general audience which included artists and movie producers. Taimina's initial lecture and following other public presentations sparked great interest in this new tactile way of exploring concepts of hyperbolic geometry, making this advanced topic accessible to wide audiences. Originally creating purely mathematical models, Taimina soon became popular as a fiber artist and public presenter for general audiences of ages five and up. In June 2005, her work was first shown as art in an exhibition "Not The Knitting You Know" at Eleven Eleven Sculpture Space, an art gallery in Washington, D.C. Since then she has participated regularly in various shows in galleries in US, UK, Latvia, Italy, Belgium, Ireland. Her artwork is in the collections of several private collectors, colleges and universities, and has been included in the American Mathematical Model Collection of the Smithsonian Museum, Cooper–Hewitt, National Design Museum, and Institute Henri Poincare.

Her work has received wide interest in media. It has been written about in 'Knit Theory' in Discover magazine. The Times Margaret Wertheim interviewed Daina Taimina and David Henderson for Cabinet Magazine. Later, the Institute For Figuring published a brochure "A Field Guide to Hyperbolic Space". In 2005 the IFF decided that to incorporate Taimina's ideas and approach of explaining hyperbolic space in their mission of popularizing Mathematics, and curate an exhibition at Machine Project gallery, which was the subject of a piece in the Los Angeles Times. Taimina's way of exploring hyperbolic space via crochet and connections with nature, combatting math phobia, was adapted by Margaret Wertheim in her talks and became highly successful in the IFF-curated Hyperbolic Crochet Coral Reef project.

Taimina's book "Crocheting Adventures with Hyperbolic Planes" won the 2009 Bookseller/Diagram Prize for Oddest Title of the Year.

Achievement

Taimina won the 2012 Euler Book Prize of the Mathematical Association of America.

Contribution

Inspired by William Thurston's paper creations back in the 1960s, she thought if something can be made out of paper, it can also be crocheted, so she made her first crocheted hyperbolic planes in June 1997 by increasing stitches in constant ratio after every two stitches she did an increase by one stitch. The number of stitches in each row grew exponentially, so after finishing her first small, very ruffled one she realized that to explore the hyperbolic plane she has to change the ratio of increase. For classroom use the best is to use the ratio 12:13 it means to increase one stitch after every 12 single crochet stitches.

"Seven Shades of Purple"

Autumn (Thanksgiving)

“Manifold II (in tree) in memory of Bill Thurston"

"Manifold II (Bird's Eye View) in memory of Bill Thurston”

“Land and the Sea”

Chapter 23

Karen Keskulla Uhlenbeck

Pioneer in mathematical analysis

Born : August 24, 1942, Cleveland, Ohio
Nationality : American

Karen Keskulla Uhlenbeck is a professor and Sid W. Richardson Foundation Regents Chairholder in the Department of Mathematics at The University of Texas in Austin.

Education and Career

Karen Keskulla Uhlenbeck was born in Cleveland, Ohio, on August 24, 1942, to Arnold Edward Keskulla, an engineer, and Carolyn Windeler Keskulla, an artist. When Uhlenbeck was in third grade, the family moved to New Jersey. Everything interested her as a child, but she felt that girls were discouraged from exploring many activities. In high school, she read American physicist George Gamow's books on Physics and English astronomer Fred Hoyle's books on cosmology, which her father brought home from the public library. When Uhlenbeck entered the University of Michigan, she found Mathematics a broad and intellectually stimulating subject. After earning her B.Sc. degree in 1964, she became a National Science Foundation Graduate Fellow, pursuing graduate study in Mathematics at Brandeis University. In 1965, she married Olke Cornelis Uhlenbeck, a biophysicist; they later divorced.

Uhlenbeck received her Ph.D. in Mathematics from Brandeis in 1968 with a thesis on the calculus of variations. Her first teaching position was at the Massachusetts Institute of Technology in 1968. The following year she moved to Berkeley, California, where she was a lecturer in Mathematics at the University of California. There she studied general relativity and the geometry of space-time, and worked on elliptic regularity in systems of partial differential equations.

In 1971, Uhlenbeck became an assistant professor at the University of Illinois at Urbana-Champaign. In 1974, she was awarded a fellowship from the Sloan Foundation that lasted until 1976, and she then went to Northwestern University as a visiting associate professor. She taught at the University of Illinois in Chicago from 1977 to 1983, first as associate professor and then professor, and in 1979 she was the Chancellor's Distinguished Visiting Professor at the University of California, Berkeley. An Albert Einstein Fellowship enabled her to pursue her research as a member of the Institute for Advanced Studies at Princeton University from 1979 to 1980.

Research

She participates or has participated in research in the fields of geometric partial differential equations, the calculus of variations, gauge theory, topological quantum field theory, and integrable systems.

Works and Awards

- She published more than a dozen articles in Mathematics journals during the 1970s and was named to the editorial board of the **Journal of Differential Geometry** in 1979 and the **Illinois Journal of Mathematics** in 1980.
- In 1983, Uhlenbeck was selected by the John D. and Catherine T. MacArthur Foundation of Chicago to receive one of its five-year fellowship grants.
- Uhlenbeck and other researchers viewed instantons as somewhat analogous to soap films. Seeking a better understanding of these particles, they studied soap films to learn about the properties of surfaces.
- As soap films provide a model for the behaviour of surfaces in three-dimensions, instantons provide analogous models for, zth behaviour of surfaces in four-dimensional space-time.
- Uhlenbeck cowrote a book on this subject, **Instantons and 4-Manifold Topology**, which was published in 1984.
- She was elected to the American Academy of Arts and Sciences in 1985 and to the National Academy of Sciences in 1986.
- In 1988, she received the Alumni Achievement award from Brandeis University, an honourary doctor of science degree from Knox College, and

was named one of America's 100 most important women by **Ladies' Home Journal**.

- She believes that Mathematics is always challenging and never boring, and she has expressed the hope that one of her accomplishments as a teacher has been communicating this to her students. "I sometimes feel the need to apologize for being a mathematician, but no apology is needed,"
- She told The Alcalde Magazine. "Whenever I get a free week and start doing Mathematics, I can't believe how much fun it is. I'm like a 12-year-old boy with a new train set."
- "Even when I had had my Ph.D. for five years," she said, "I was still struggling with whether I should become a mathematician. I never saw myself very clearly."
- In 1982, she received a MacArthur Fellowship, the so-called "genius" award.
- In 1986, at age 44, Uhlenbeck became the first woman mathematician elected to the National Academy of Sciences.
- Since then, only two other women mathematicians have been elected to the Academy—both in the last two years—out of a total of 111 mathematicians who are current members.
- In announcing her election to the National Academy of Sciences, the Academy said: **"Uhlenbeck has been the foremost pioneer in developing and applying hard analytic methods in the calculus of variations and nonlinear partial differential equations to the study of the mathematical structure of the gauge field theories of present-day fundamental Physics."**
- In 1995, Uhlenbeck received the Common Wealth Award for Science and Technology from Sigma Xi, the national research honourary society, and, in 1984, was named Alumna of the Year by the U-M.
- In 1988 she was selected to be a Noether Lecturer.
- In 2000, she became a recipient of the National Medal of Science.
- In 2007 she won the American Mathematical Society Steele Prize and was awarded an honourary doctorate from Harvard University. She has also been a MacArthur Fellow.
- In 2012 she became a fellow of the American Mathematical Society.

Uhlenbeck says she entered Mathematics believing she could simply work by herself and hide from the pressures women faced in academia. She ultimately forged a number of collaborations that helped her work. She also learned that she truly enjoyed mentoring students. She is especially proud of her role as one of the founders of the Park City Mathematics Institute at the Institute for Advanced Study at Princeton.

Under the auspices of this Institute, Uhlenbeck has organized a mentoring program for women mathematicians. Over a two-week period, the women participate in seminars, working problem groups, and mentoring and networking sessions. They also have an opportunity to meet and converse with mathematicians in residence at the Institute for Advanced Study. Karen Uhlenbeck has blended her unique interests, a curiosity for Mathematics and inquiry, and a doggedness to surpass the social hurdles she has encountered to create a successful career as a mentor and scientist.

Chapter 24

Cauchy

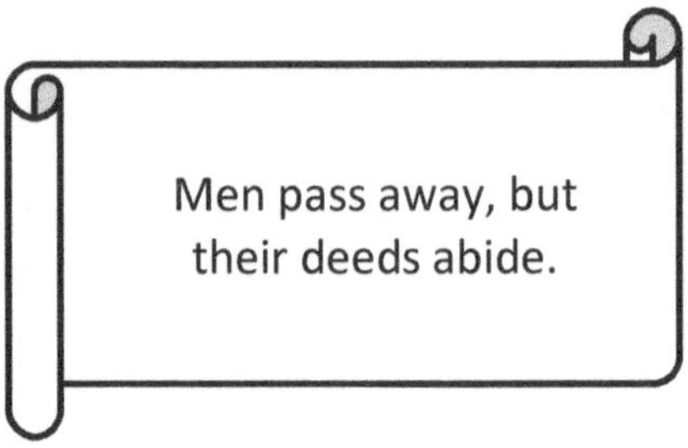

Born : August 21, 1789, Paris
Died : May 23, 1857, Sceaux
Nationality : French

Youth and education

Cauchy was the son of Louis Francois Cauchy (1760–1848) and Marie-Madeleine Desestre. Cauchy had two brothers, Alexandre Laurent Cauchy; who became a judge of the court of cassation in 1849; and Eugene Francois Cauchy, a publicist who also wrote several mathematical works. He laid the foundation for modern day analysis and did many significant works in the field of Mathematics.

The famous mathematician Lagrange was also a friend of the Cauchy family. On Lagrange's advice, Augustin-Louis was enrolled in the Ecole Centrale du Pantheon, the best secondary school of Paris at that time, in the fall of 1802. In spite of these successes, Augustin-Louis chose an engineering career, and prepared himself for the entrance examination to the Ecole Polytechnique.

In 1805 he placed second out of 293 applicants on this exam, and he was admitted. One of the main purposes of this school was to give future civil and military engineers a high-level scientific and mathematical education.

Junior Engineer

After finishing school in 1810, Cauchy accepted a job as a junior engineer for three years in Cherbourg, where Napoleon intended to build a naval base.

In September 1812, now 23 years old, after becoming ill from overwork, Cauchy returned to Paris. Another reason for his return to the capital was that he was losing his interest in his engineering job, being more and more attracted to the abstract beauty of Mathematics; in Paris. The next three years Augustin-Louis was mainly on unpaid sick leave, and spent his time quite fruitfully, working on Mathematics.

He attempted admission to the First Class of the Institute France but failed on three different occasions between 1813 and 1815. In 1815 Napoleon was defeated at Waterloo, and the newly installed Bourbon king Louis XVIII took the restoration in hand.

Rising Mathematical Star

In November 1815, Louis Poinsot, who was an associate professor at the Ecole Polytechnique, asked to be exempted from his teaching duties for health reasons. Cauchy was by then a rising mathematical star, who certainly merited a professorship. One of his great successes at that time was the proof of Fermat's polygonal number theorem.

He finally quit his engineering job, and received a one-year contract for teaching Mathematics to second-year students of the Ecole Polytechnique. In 1816, this Bonapartist, non-religious school was reorganized, and several liberal professors were fired; the reactionary Cauchy was promoted to full professor.

In 1824 Louis XVIII died, and was succeeded by his even more reactionary brother Charles X. During these years Cauchy was highly productive, and published one important mathematical treatise after another. Cauchy remained a professor at the University until his death at the age of 67. He received the Last Rites and died at 4 a.m. on May 23, 1857. His name is one of the 72 names inscribed on the Eiffel Tower.

Contributions

- ***Wave theory, mechanics, elasticity***

He also contributed significant research in mechanics, substituting the notion of the continuity of geometrical displacements for the principle of the continuity of matter. He introduced a 3×3 symmetric matrix of numbers that is now known as the Cauchy stress tensor. In elasticity, he originated the theory of stress, and his results are nearly as valuable as those of Simeon Poisson.

- ***Number theory***

Other significant contributions include being the first to prove the Fermat polygonal number theorem.

✶ *Complex functions*

Cauchy is most famous for his single-handed development of complex function theory. The first pivotal theorem proved by Cauchy, now known as *Cauchy's integral theorem*, was the following:

$$\oint_C f(z)dz = 0 ,$$

where *f*(*z*) is a complex-valued function holomorphic on and within the non-self-intersecting closed curve *C* lying in the complex plane. The *contour integral* is taken along the contour *C*. The rudiments of this theorem can already be found in a paper that the 24-year-old Cauchy presented to the Academie des Sciences on August 11, 1814. In full form the theorem was given in 1825. The 1825 paper is seen by many as Cauchy's most important contribution to Mathematics.

In 1826 Cauchy gave a formal definition of a residue of a function. This concept regards functions that have poles—isolated singularities, i.e., points where a function goes to positive or negative infinity. If the complex-valued function *f*(*z*) can be expanded in the neighborhood of a singularity *a* as

$$f\left(z) = \phi(z) + \frac{B_1}{z-a} + \ldots + \frac{B_n}{(z-a)^n}, B_i, z, a \in C, \right)$$

where φ(*z*) is analytic (i.e., well-behaved without singularities), then *f* is said to have a pole of order *n* in the point *a*. If $n = 1$, the pole is called simple. The coefficient B_1 is called by Cauchy the residue of function *f* at *a*. If *f* is non-singular at *a* then the residue of *f* is zero at *a*. Clearly the residue is in the case of a simple pole equal to,

$$\operatorname*{Res}_{z=ak} f(z) = \lim_{z \to a} (z-a)\, f\,(z),$$

where we replaced B_1 by the modern notation of the residue.

In 1831, while in Turin, Cauchy submitted two papers to the Academy of Sciences of Turin. In the first he proposed the formula now known as Cauchy's integral formula,

$$f(a) = \frac{1}{2\pi i} \oint_C \frac{f(z)}{z-a} dz,$$

where *f*(*z*) is analytic on *C* and within the region bounded by the contour *C* and the complex number *a* is somewhere in this region. The contour integral is taken counter-clockwise. Clearly, the integrand has a simple pole at $z = a$. In the second paper he presented the residue theorem,

$$\frac{1}{2\pi i} \oint_C f(z)dz = \sum_{k=1}^{n} \operatorname{Res} f(z)),$$

where the sum is over all the *n* poles of *f*(*z*) on and within the contour *C*. These results of Cauchy's still form the core of complex function theory as it is taught today to physicists and electrical engineers.

In his book *Cours d'Analyse* Cauchy stressed the importance of rigor in analysis. The book is frequently noted as being the first place that inequalities, and $\delta-\in$ arguments were introduced into Calculus. Here Cauchy defined continuity as follows: *The function f(x) is continuous with respect to x between the given limits if, between these limits, an infinitely small increment in the variable always produces an infinitely small increment in the function itself.*

✶ ***Taylor's theorem***

He was the first to prove Taylor's theorem rigorously, establishing his well-known form of the remainder. He wrote a textbook for his students at the Ecole Polytechnique in which he developed the basic theorems of mathematical analysis as rigorously as possible.

In this book he gave the necessary and sufficient condition for the existence of a limit in the form that is still taught. Also Cauchy's well-known test for absolute convergence stems from this book: Cauchy condensation test. In 1829 he defined for the first time a complex function of a complex variable in another textbook.

✶ ***Argument principle, stability***

In a paper published in 1855, two years before Cauchy's death, he discussed some theorems, one of which is similar to the "Argument Principle" in many modern textbooks on complex analysis. In modern control theory textbooks, the Cauchy argument principle is quite frequently used to derive the Nyquist stability criterion, which can be used to predict the stability of negative feedback amplifier and negative feedback control systems. Thus Cauchy's work has a strong impact on both pure Mathematics and practical engineering.

Cauchy was very productive, in number of papers second only to Leonhard Euler. It took almost a century to collect all his writings into 27 large volumes.

Major Works

Augustin-Louis Cauchy is best known for single-handedly developing Complex analysis, traditionally known as the theory of functions of a complex variable. This branch of mathematical analysis investigates functions of complex numbers, and is useful in many branches of Mathematics, as well as Physics.

Personal Life and Legacy

- In 1818, Cauchy married Aloise de Bure, member of a family of publishers and booksellers who published majority of his works. The couple had two daughters: Marie Francoise Alicia and Marie Mathilde Augustin.
- Cauchy died on May 23, 1857.
- No other mathematician except Leonhard Euler left behind a bigger body of work than Cauchy.

Later Life

There is not much to Cauchy's personal life except that he got married at the age of 28 to Aloise de Bure and had two daughters. In July 1830, France went through a revolution when Charles X was succeeded by Louis-Phillippe resulting in many riots in which the students of Polytechnique took part actively. These events turned around Cauchy's situation completely which resulted in a halt to his mathematical work.

Cauchy left France for some time but he had to take a decision of whether he was going to take sides with the new regime or not. He decided against it which cost him all his posts in Paris except for his membership of the Académie des Sciences. This did not stop him from pursuing further mathematical work so he went to Turin, Italy where he got an offer from the King of Sardinia for chair of Physics which he accepted happily. He remained at the post for a year before he left for Prague to tutor Duke of Bordeaux till 1838. Their relationship as teacher and student was not great but Cauchy still tried his best to teach his pupil the most he could. Cauchy did not gain his academic positions again however tried to remain in touch with his roots till his death in 1857.

Thiravialakshmi T

1st Year B.Ed Student

Chapter 25

Arthur Cayley

Projective Geometry is all Geometry

Born : **August 16, 1821, Richmond, Surrey**
Died : **January 26, 1895, Cambridge**
Nationality : **British**

Arthur Cayley was a British mathematician who helped founding the British school of pure Mathematics. He is most famous for developing the algebra of matrices and non-Euclidean and n-dimensional geometry. Cayley worked primarily with algebra and published several papers on Mathematics. Even as a child, Cayley loved solving mathematical problems. He enjoyed the subject and was also an avid reader. He pursued his studies at Trinity College, Cambridge and entered the field of Mathematics. He was also fluent in French, German, Greek and Italian.

Childhood and Education

Cayley was born in Richmond, England on 16th August 1821 to Henry Cayley and Maria Antonia Doughty. His brother Charles Bagot Cayley was a linguist. His father, a merchant, settled with his family in Saint Petersburg. For the first eight years, Cayley grew up in Germany. In 1829 they moved back to England and settled in Blackheath, near London.

Cayley was sent to a private school for four years, and even at his young age, he had an affinity for Mathematics. At the age of fourteen, he was sent to King's College School. His genius in Mathematics was duly noted and his teacher told him to pursue Mathematics instead of falling into his father's footsteps. In 1838, at the tender age of seventeen, Cayley began learning at Trinity College, Cambridge. He finished his undergraduate course with honours, in 1842.

He was awarded the **'Senior Wrangler'** in **'Mathematical Tripos'** and first place in the competition for the **'Smith Prizes'**. After this, he continued his studies by taking up an M.A. degree and won a fellowship. He resided in Cambridge after winning a Fellowship. During this period, he gave lectures on Mathematics. His first contribution was made in 1841, to the 'Cambridge Mathematical Journal' established by Gregory and Robert Leslie Ellis. He submitted three papers on subjects based on reading the **'Mecanique analytique of Lagrange'** and some of the works of 'Laplace'. His main works, however, were the twenty-eight memoirs to the **'Mathematical Journal'**.

Work

At the end of seven years, Cayley ended his fellowship at Cambridge and had to choose a profession. Cayley opted for law and joined Lincoln's Inn, London. As a lawyer, he specialized in conveyancing. During these fourteen years, Cayley wrote about 200-300 papers.

In 1863, he accepted the Sadleirian professorship in Mathematics at Cambridge. He devoted himself to mathematical research and gave up on his career in law. When it came to Mathematics, Cayley was known to be an individualist. His skill was formidable and it was said that he had an intuitive understanding of mathematical theories. Cayley made valuable contributions to algebraic theory of curves and surfaces, graph theory, group theory, linear algebra, combinatorics and elliptic functions.

In geometry, however, his work was based on analytic geometry. In 1859, Cayley was the first person to realize that **Euclidean geometry** was a special case of projective geometry and ten years later, his projective metric helped people understand the relationship between the different types of non-Euclidean geometries.

Cayley served as the **President of the BMS (British Association for the Advancement of Science),** the Cambridge Philosophical Society, the London Mathematical Society and the Royal Astronomical Society. Cayley wrote over a thousand papers in his lifetime. In **1876**, Cayley published a book called **'An Elementary Treatise on Elliptic Functions'.** This was the only book he wrote and it was based on his studies of the Karl Jacobi's point of view.

Contribution

Cayley made important contributions to the algebraic theory of curves and surfaces, group theory, linear algebra, graph theory, combinatorics, and elliptic functions. He formalized the theory of matrices. Among Cayley's most important papers were his series of 10 **"Memoirs on Quantics" (1854–78).** A quantic, known today as an algebraic form, is a polynomial with the same total degree for each term. Alongside work produced by his friend James Joseph Sylvester, Cayley's study of various properties of forms that are unchanged (invariant) under some transformation, such as rotating or translating the coordinate axes, established a branch of algebra known as invariant theory.

Death

Due to natural causes, Cayley passed away on 26th January 1895.

Honours

The honours that were awarded to Arthur Cayley are:

- Fellow of the Royal Society, 1852
- Royal Society Royal Medal, 1859
- Fellow of the Royal Society of Edinburgh, 1865
- Royal Society Copley Medal, 1882
- LMS De Morgan Medal, 1884

Legacy

Cayley proposed many theorems and mathematical papers. Some of his famous works are:

- Cayley's theorem
- Cayley's Ω process
- Cayley–Hamilton theorem
- Cayley algebra
- Cayley–Purser algorithm

Arthur Cayley Timeline

- 1821:Arthur Cayley was born on 16 August in Richmond, Surrey, UK
- 1838:Joined Trinity College, Cambridge
- 1841:He submitted three papers to the 'Cambridge Mathematical Journal'; his first publication
- 1859:He was the first person to realize that Euclidean geometry was a special case of projective geometry
- 1863:He accepted the Sadleirian professorship in Mathematics at Cambridge
- 1876:He published a book called, 'An Elementary Treatise on Elliptic Functions'
- 1895:Passed away on 26th January in Cambridge, England.

Chapter 26

Johaan Bernoulli

Wisdom is found only in truth.

Born : August 6, 1667, Basel
Died : January 1, 1748, Basel
Nationality : Swiss

Childhood and Early Life

Johann Bernoulli was born to Nicolaus Bernoulli and Margeretha Schonauer. Nicolaus wanted him to study business to continue on the family spice trade business. In 1683, he attended 'Basel University' to study medicine, contrary to his father's wishes. However, he ended up not liking medicine. He studied Mathematics with his older brother Jacob outside of school. Twelve years his senior, Jacob was involved in the study of a newer field called calculus and he shared his findings with Johann.

The brothers wrote letters of their discoveries to each other, coining the term **'integral'** as it relates to math. However, jealousy began to tear the two of them apart as Johann left his brother out of certain communications with Leibniz, a famous mathematician of the day. Johann was betrayed when one of the first publications of the material he and his brother had been working on was released by his pupil, Guillaume de l'Hopital.

Stung but resilient, he began to teach Mathematics as the 'University of Basel', taking over his deceased brother's position. In the years that followed, he released many mathematical papers and supported Leibniz in his debate against Newton and his theory of gravity. He let jealousy sabotage a family relationship once more as he plagiarized his son's mathematical paper. Johann Bernoulli was honoured with many academy memberships during his lifetime.

Career

He began teaching at University of Groningen in 1694 but was asked to return to his hometown of Basel by his father-in-law after nine years of teaching there.

In 1696, controversy struck when he agreed to tutor Guillaume de l'Hopital. l'Hopital published a textbook on calculus creating 'l'Hopital's Rule', stealing all of Bernoulli's idea – but there was nothing he could do, as he had signed a contract saying l'Hopital could use his discoveries.

A year later, in 1697, he proved the **'Sophomore's dream'**, an important mathematical theorem still used to this day. He proposed the **'Brachistorchone'** problem, in 1969, that was later solved by Newton, Leibniz, l'Hopital, and his brother Jacob. The object was to form a wire in a way that would let a ball traveled down it in the fastest manner possible.

In 1705, he learned that Jacob had passed away due to tuberculosis. He was able take his brother's old position of Mathematics professor at 'Basel University'. He participated in the Newton-Leibniz debate, in 1713, arguing that Leibniz deserved credit for the discovery of calculus because his method was more effective in solving certain problems.

Later, he argued against Newton's theory of gravity and delayed the acceptance of Newton's theory throughout Europe. In 1726, he published a paper based on a competition sponsored by the **'French Academy of Science'** two years earlier. Although he had not won the competition, his paper on motion was considered valid.

In 1738, he found a rival in his own son, Daniel, as the two of them attempted to publish a paper of hydrodynamics in the same year. The father dated his paper two years prior to the actual publication date to combat the competition

Personal Life and Legacy

- In 1694, he married Dorothea Falkner. She was the daughter of an Alderman of Basel and together they had three children Nicolaus II, Daniel, and Johann II.
- It is reported that he kicked his son Daniel out of their home after the son took home a mathematical prize in the competition they both had entered.
- He became known as the 'Archimedes of his age' and enjoyed the title so much he had it inscribed on his tombstone upon his death in 1748.
- He published many acclaimed papers and mathematical text throughout his life and also completed an autobiography.

- He was famous for teaching Leonhard Euler, who would publish over 900 works about Mathematics in his lifetime.

Awards and Achievements

- He was awarded membership to the 'Academy of Science' in Paris, Berlin, St. Petersburg, London, and many others.

Death

Johann Bernoulli died at the age of 80 on 1st January 1748.

Chapter 27

Jean Baptiste Joseph Fourier

Our mission is to move forward

Born : March 21, 1768 AD, Auxerre
Died : May 16, 1830 AD, Paris
Nationality : French

A French mathematician and physicist, Joseph Fourier is renowned for showing how the conduction of heat in solid bodies could be analyzed in terms of infinite mathematical series, which is fondly called the **'Fourier Series'.** Not only is this prominent mathematician and physicist remembered for initiating the investigation of Fourier series, but is paid regard to for the application of the same to the problems of heat transfer and vibrations.

It is interesting to note that Fourier's theory was the first accurate theory on heat diffusion. Originally trained for priesthood, fate had something else in store for this great man. A close comrade of Napoleon, Joseph Fourier played an important role in the Egyptian expeditions as well. Joseph Fourier is also credited with the discovery of the greenhouse effect. Know more about this tremendously talented mathematician and physicist, read through the following lines.

Childhood and Early Life

Joseph Fourier was born in a modest family at Auxerre, France. Orphaned at an early age, Joseph acquired his initial education from Pallais's school, which was run by the music teacher from the cathedral. Thereafter, he proceeded to the Ecole Royale Militaire of Auxerre. While he showed an aptitude and flair for literature, by the age of thirteen, this was overshadowed by Mathematics, a subject he found himself really interested in. Joseph completed a study of the six volumes of Bezout's Cours de mathematiques when he was fourteen and in the following year, received the first prize for his study of Bossut's Mécanique en général.

In the year 1787, Fourier took admission in the Benedictine abbey of St Benoit-sur-Loire, with a motive to train himself for priesthood. However, simultaneously, he submitted a paper on algebra to Montucla in Paris and even wrote a letter to Bonard stating his desire to make a major influence in Mathematics. This move suggested Fourier's uncertainty as to whether he really wanted to pursue the training in priesthood or not. In 1789, Fourier left for Paris, where he read a paper on algebraic equations at the Académie Royale des Sciences.

The following year, Fourier took up the position of a tutor at the Benedictine college, École Royale Militaire of Auxerre, where he had studied. As if Fourier's uncertainty to follow a religious life or one of mathematical research was not enough, the conflict became worse when he got involved in politics and joined the local Revolutionary Committee. Joseph Fourier returned to Auxerre and continued to work on the revolutionary committee as well as teaching at the college. In 1794, Fourier was imprisoned for a brief duration but was freed soon. A year later, he was appointed to study at the École Normale in Paris, a teacher-training school set up for training teachers.

Fourier, as expected, turned out to be the brightest student of the institution. Fourier was taught by the best teachers of that time, including Lagrange, Laplace and Monge. Later, Fourier began teaching at the Collège de France. He maintained good relations with his teachers and thus with their help, started off on his journey of mathematical research. Fourier progressed steadily, being appointed to a position at the École Centrale des Travaux Publiques or École Polytechnique as it was renamed later. However, ramifications of his earlier arrest came to the forefront as Fourier was re-arrested and imprisoned. Nevertheless, this did not last long as he was soon released.

His Works

In 1822, Fourier presented his papers on the heat flow in Théorie analytique de la chaleur (The Analytic Theory of Heat). Based on Newton's law of cooling, Fourier interpreted that the flow of heat between two adjacent molecules is directly proportional to the extremely small difference of their temperatures. There were three aspects of the work, one mathematical and two physical.

In Mathematics, Fourier argued that any function of a variable, whether continuous or discontinuous, could be expanded in a series of sines of multiples of the variable. Though this was an incorrect judgement made by him, Fourier's observation that some discontinuous functions are the sum of infinite series was a huge revelation. One amongst the two physical contributions mentioned in the book was the concept of dimensional homogeneity in equations. According to it, an equation could be formally correct only if the dimensions matched on either side of the equality. The other significant physical contribution was Fourier's proposal of his partial differential equation for conductive diffusion of heat. Till date, this equation is taught to every student of mathematical Physics

In addition to the above, Fourier left an unfinished work on determinate equations which was completed and published by Claude-Louis Navier in 1831. This work included a demonstration of Fourier's theorem on the position of the roots of an algebraic equation. Fourier, alongside his mathematical ventures, introduced the concept of greenhouse for the first time ever in history.

According to his calculation, Earth, taking in consideration its size and distance from the Sun, should actually be much colder than what it actually is, if warmed only by the effects of incoming solar radiation. He concluded that a large portion of the additional heat that the planet received was due to the interstellar radiation.

His concept that the Earth's atmosphere acted as an insulator of a kind was the first ever suggestion of what it is today known as the greenhouse effect. Referring to an experiment conducted by de Saussure, Fourier suggested that gases in the atmosphere could form a stable barrier like the glass panes. This laid the foundation for what we today know as greenhouse effect.

Example of a *Fourier Series*

In Fourier's most famous work, *Memoir on the Propagation of Heat in Solid Bodies*, one finds the following mathematical expression, which is a simple example of what is now called a *Fourier series*:

$$\varphi(y) = a \cos\frac{\pi y}{2} + a' \cos 3\frac{\pi y}{2} + a'' \cos 5\frac{\pi y}{2} + \ldots.$$

The dots at the end of the series mean that the terms of the series continue indefinitely in the same pattern shown. Although other mathematicians had used trigonometric series, Fourier appears to have been the first to realize that they could be used to represent functions of an arbitrary nature, including discontinuous functions.

Contributions in Algebra

Fourier left an unfinished work on determinate equations which was edited by Claude-Louis Navier and published in 1831. This work contains much original matter—in particular, there is a demonstration of Fourier's theorem on the position of the roots of an algebraic equation. Joseph Louis Lagrange had shown how the roots of an algebraic equation might be separated by means of another equation whose roots were the squares of the differences of the roots of the original equation. François Budan, in 1807 and 1811, had enunciated the theorem generally known by the name of Fourier, but the demonstration was not altogether satisfactory. Fourier's proof is the same as that usually given in textbooks on the theory of equations. The final solution of the problem was given in 1829 by Jacques Charles François Sturm.

Death and Legacy

Fourier's health started deteriorating in 1830. While he had experienced attacks of aneurism of the heart when he was in Egypt and Grenoble, it was in Paris that the problem of suffocation became worse. A fall from the stairs on May 4, 1830, further aggravated the malady. Few days later, on May 16, 1830 Fourier breathed his last. Joseph Fourier is one amongst the 72 others, to have his name inscribed on the Eiffel Tower.

Joseph Fourier Timeline

1768	Joseph Fourier was born
1782	Received the first prize for his study of Bossut's Mécanique en général
1787	Took admission in the Benedictine abbey of St Benoit-sur-Loire to train himself for priesthood
1789	Left for Paris
1794	Fourier was imprisoned
1795	Appointed to study at the École Normale in Paris; resumed his teaching at the École Polytechnique
1797	Succeeded Lagrange and was appointed to the chair of analysis and mechanics
1798	Joined Napoleon's army in its invasion of Egypt as scientific advisor
1801	Returned to France and resumed his post as Professor of Analysis at the École Polytechnique.
1816	Presented the paper "On the Propagation of Heat in Solid Bodies" to the Paris Institute; Moved to England
1822	Returned to France in 1822; Presented papers on the heat flow in Théorie analytique de la chaleur (The Analytic Theory of Heat)
1830	Joseph Fourier passed away

Josephine Prabha F

1st Year B.Ed Student

Chapter 28

Ernst Eduard Kummer

Father of Modern Arithmetic

Born	:	**January 29, 1810, Żary, Poland**
Died	:	**May 14, 1893, Province of Brandenburg**
Nationality	:	**German**

Ernst Eduard Kummer (29 January 1810–14 May 1893) was a German mathematician. Skilled in applied Mathematics, Kummer trained German army officers in ballistics; afterwards, he taught for 10 years in a gymnasium, the German equivalent of high school, where he inspired the mathematical career of Leopold Kronecker.

Ernst Kummer was born in Sorau, Brandenburg to his father CarlGotthelf Kummer, a physician. However his father died when Ernst was only three years old and Ernst and his elder brother Karl were brought up by their mother. Ernst received privatecoaching before entering the Gymnasium in Sorau when he was nine years old. In 1828 Kummer entered the University of Halle to study Protestant theology. But fortunately for the good of Mathematics, Kummer received Mathematics teaching as part of his degree to provide a proper foundation to the study of philosophy. Soon the study of Mathematics should become his main subject, although at this stage he still saw it as leading to a later study of philosophy.

In 1831 Kummer was awarded a prize for a mathematical essay and in the same year he was awarded his certificate to enable him to teach in schools. Moreover, on the strength of his prize winning essay, he was awarded a doctorate. After a probationary year at the Gymnasium in Sorau where he had been educated, he was appointed to a teaching post at the Gymnasium in Liegnitz, now Legnica in Poland that he held for the next 10 years. Some of his pupils there had great ability, most prominent among them Leopold Kronecker, and conversely they were extremely fortunate to find a school teacher of Kummer's quality and ability to inspire.

Early life

Kummer was born in Sorau, Brandenburg (then part of Prussia). He was awarded a Ph.D from the University of Halle in 1831 for writing a prize-winning mathematical essay (De cosinuum ET sinus potestatibus secundum cosines ET sinus arcuum multiplicium evolvendis), which was eventually published a year later.

Kummer was married in 1840 to Ottilie Mendelssohn, daughter of Nathan Mendelssohn and Henrietta Itzig. Ottilie was a cousin of Felix and his sister Rebecca Mendelssohn Bartholdi, the wife of the mathematician Peter Gustav Lejeune Dirichlet. His second wife (which he married soon after the death of Ottilie in 1848), Bertha Cauer, was a maternal cousin of Ottilie. Overall, he had 13 children. His daughter Marie married the mathematician Hermann Schwarz. Kummer retired from teaching and from Mathematics in 1890 and died three years later in Berlin.

Contributions and Achievements

Kummer made several contributions and achievements in different areas.

- He codified some of the relations between differently known as contiguity relations. The Kummer surface results from taking the quotient of a two-dimensional abelian variety by the cyclic group $\{1, -1\}$ (an early orbifold: it has 16 singular points, and its geometry was intensively studied in the nineteenth century).
- Kummer also proved Fermat's last theorem for a considerable class of prime exponents. His methods were closer, perhaps, to p-adic ones than to ideal theory as understood later, though the term 'ideal' arose here.
- He studied what were later called Kummer extensions of fields: that is, extensions generated by adjoining an *n*th root to a field already containing a primitive *n*th root of unity. This is a significant extension of the theory of quadratic extensions, and the genus theory of quadratic forms as such, it is still foundational for class field theory.

- One of his major contributions is the introduction of ideal numbers, which are defined as a special subgroup of a ring, extended the fundamental theorem of arithmetic to complex number fields.
- He also discovered the fourth order surface based on the singular surface of the quadratic line complex. This Kummer surface has 16 isolated conical double points and 16 singular tangent planes.
- During Kummer‘s first period of Mathematics he worked on function theory.
- He extended Gauss‘s work onhypergeometric series, giving developments that are useful in the theory of differential equations.
- He was the first to compute the monodromy groups of these series.
- In 1839, although still a school teacher, Kummer was elected to the Berlin Academy on Dirichlet‘s recommendation.
- In 1842, Kummer became professor of Mathematics at the University of Breslau (now Wrocław, Poland).
- In 1855 he succeeded Peter Gustav Lejeune Dirichlet as professor of Mathematics at the University of Berlin, at the same time also becoming a professor at the Berlin War College.
- In1843, Kummer showed Dirichlet an attempted proof of Fermat’s last theorem. But, Dirichletfound an error, and Kummer continued his search and developed the concept of so-called ideal numbers. In number theory an ideal number is an algebraic integer which represents an ideal in the ring of integers of a number field. An ideal is a special subset of an algebraic ring. Ideals generalize certain subsets of the integers, such as the even numbers or the multiples of 3. Addition and subtraction of even numbers preserves evenness, and multiplying an even number by any other integer results in another even number.
- Using the concept of ideal numbers, Kummer proved the insolubility of the Fermat relation for all but a small group of primes, and he thus laid the foundation for an eventual complete proof of Fermat’s last theorem. For his great advance, the French Academy of Sciences awarded him its Grand Prize in 1857. The ideal numbers have made possible new developments in the arithmeticof algebraic numbers.
- In 1861, Germany’s first seminar in pure Mathematics was established at Berlin on the recommendation of Kummer and Karl Weierstrass. It soon attracted gifted young mathematicians form throughout the world, including many graduate students.
- Kummer’s Berlin lectures, always carefully prepared, covered analytic geometry, mechanics, the theory ofsurfaces, and number theory.

Publications

- Kummer, Ernst Eduard (1975), Weil, André, ed., Collected papers. Volume 1: Contributions to Number Theory, Berlin, New York: Springer-Verlag, ISBN 978-0-387-06835-0, MR 0465760
- Kummer, Ernst Eduard (1975), Weil, André, ed., Collected papers. Volume II: Function theory, geometry and miscellaneous, Berlin, New York: Springer-Verlag, ISBN 978-3-540-06836-5, MR 0465761

For the rest of his life, Kummer was devoted to geometry. He applied himself with unbroken productivity to ray systems and also considered ballistic problems. By that time, also discovered the fourth order surface, now named after him, based on the singular surface of the quadratic line complex. The Kummer surface has 16 isolated conical double points and 16 singular tangentplanes and was published in 1864. He retired at his own request in 1883 to spent the last yearsof his life in quiet retirement until his death in 1893.

Chapter 29

Sir William Rowan Hamilton

A commemorative coin was issued by the Central Bank of Ireland in his honour.

Born : August 4, 1805, Dublin
Died : September 2, 1865, Dublin
Nationality : Irish

Early life

Hamilton was the fourth of nine children born to Sarah Hutton and Archibald Hamilton, who lived in Dublin at 38 Dominick Street. Hamilton's father who was from Dunboyne worked as a solicitor. By the age of three, Hamilton had been sent to live with his uncle James Hamilton, a graduate of Trinity College who ran a school in Talbots Castle in Trim Co. Meath. His uncle soon discovered that Hamilton had a remarkable ability to learn languages, and from a young age, had displayed an uncanny ability to acquire them.

At the age of seven he had already made very considerable progress in Hebrew, and before he was thirteen he had acquired, under the care of his uncle (a linguist), almost as many languages as he had years of age. These included the classical and modern European languages, and Persian, Arabic, Hindustani, Sanskrit and even Marathi and Malay. He retained much of his knowledge of languages to the end of his life, often reading

Persian and Arabic in his spare time, although he had long stopped studying languages, and used them just for relaxation.

In September 1813 the American calculating prodigy Zerah Colburn was being exhibited in Dublin. Colburn was 9, a year older than Hamilton. The two were pitted against each other in a mental arithmetic contest with Colburn emerging the clear victor. In reaction to his defeat, Hamilton dedicated less time to studying languages and more time to studying Mathematics.

Hamilton was part of a small but well-regarded school of mathematicians associated with Trinity College, Dublin which he entered at age 18. He studied both classics and Mathematics, and was appointed Professor of Astronomy in 1827; prior to his graduation taking up residence at Dun sink Observatory where he spent the rest of his life.

Mathematical Studies and Career

Hamilton's mathematical studies seem to have been undertaken and carried to their full development without any assistance whatsoever, and the result is that his writings do not belong to any particular "school". Hamilton was an expert as an arithmetic calculator and he seems to have occasionally had fun in working out the result of some calculation to an enormous number of decimal places.

At the age of eight Hamilton engaged Zerah Colburn, the American "calculating boy", who was then being exhibited as a curiosity in Dublin. Two years later, aged ten, Hamilton stumbled across a Latin copy of Euclid, which he eagerly devoured; and at twelve he studied Newton's Arithmetical Universalist. This was his introduction to modern analysis. Hamilton soon began to read the Principia, and at sixteen Hamilton had mastered a great part of it, as well as some more modern works on analytical geometry and the differential calculus.

Around this time Hamilton was also preparing to enter Trinity College, Dublin, and therefore had to devote some time to classics. In mid-1822 he began a systematic study of Laplace's Mecanique Celeste. From that time Hamilton appears to have devoted himself almost wholly to Mathematics, though he always kept himself well acquainted with the progress of science both in Britain and abroad. Hamilton found an important defect in one of Laplace's demonstrations, and he was induced by a friend to write out his remarks, so that they could be shown to Dr. John Brinkley, then the first Royal Astronomer of Ireland, and an accomplished mathematician. Brinkley seems to have immediately perceived Hamilton's talents, and to have encouraged him in the kindest way.

Hamilton's career at College was perhaps unexampled. Amongst a number of extraordinary competitors, he was first in every subject and at every examination. He achieved the rare distinction of obtaining an optime both for Greek and for Physics. Hamilton might have attained many more such honours if his career as a student had not been cut short by an unprecedented event. This was Hamilton's appointment to the Andrews Professorship of Astronomy in the University of Dublin, vacated by Dr. Brinkley in 1827. The chair was not exactly offered to him, as has been sometimes asserted, but the electors, having met and talked over the subject, authorized Hamilton's personal friend to urge Hamilton to become a candidate, a step which Hamilton's modesty had prevented him from taking. Thus, when barely 22, Hamilton was established at the Dun sink Observatory, near Dublin.

Hamilton was not especially suited for the post, because although he had a profound acquaintance with theoretical astronomy, he had paid little attention to the regular work of the practical astronomer. Hamilton's time was better employed in original investigations than it would have been spent in observations made even with the best of instruments. If Hamilton had devoted himself to practical astronomy, the University of Dublin would assuredly have furnished him with instruments and an adequate staff of assistants.

Quaternions

Quaternion Plaque on Broom Bridge

The other great contribution Hamilton made to mathematical science was his discovery of quaternions in 1843. However, in 1840, Benjamin Olinde Rodrigues had already reached a result that amounted to their discovery in all but name.

Hamilton was looking for ways of extending complex numbers to higher spatial dimensions. He failed to find a useful 3-dimensional system but in working with four dimensions he created quaternions. According to Hamilton, on 16th October he was out walking along the Royal Canal in Dublin with his wife when the solution in the form of the equation $i^2 = j^2 = k^2 = ijk = -1$ suddenly occurred to him; Hamilton then promptly carved this equation using his penknife into the side of the nearby Broom Bridge (which Hamilton called Brougham Bridge). This event marks the discovery of the quaternion group.

A plaque under the bridge, was unveiled by the Taoiseach Eamon de Valera (himself a mathematician and student of quaternions), on 13th November 1958. Since 1989, the National University of Ireland, Maynooth has organized a pilgrimage; where mathematicians take a walk from Dun sink Observatory to the bridge, where no trace of the carving remains, though a stone plaque does commemorate the discovery.

The quaternion involved abandoning commutative, a radical step for the time. Not only this, but Hamilton invented the cross and dot products of vector algebra. Hamilton also described a quaternion as an ordered four-element multiple of real numbers and described the first element as the 'scalar' part, and the remaining three as the 'vector' part.

Hamilton introduced, as a method of analysis, both quaternions and biquaternions, the extension to eight dimensions by introduction of complex number coefficients. When his work was assembled in 1853, the book Lectures on Quaternions had "formed the subject of successive courses of lectures, delivered in 1848 and subsequent years, in the Halls of Trinity College, Dublin".

Hamilton confidently declared that quaternions would be found to have a powerful influence as an instrument of research. When he died, Hamilton was working on a definitive statement of quaternion science. His son William Edwin Hamilton brought the Elements of Quaternions, a hefty volume of 762 pages, to publication in 1866. As copies ran short, a second edition was prepared by Charles Jasper Joly, when the book was split into two volumes, the first appearing 1899 and the second in 1901. The subject index and footnotes in this second edition improved the Elements accessibility.

Peter Guthrie Tait among others advocated the use of Hamilton's quaternions. They were made a mandatory examination topic in Dublin, and for a while they were the only advanced Mathematics taught in some American universities. However, controversy about the use of quaternions grew in the late 19th century. Some of Hamilton's supporters opposed the growing fields of vector algebra and vector calculus (from developers like Oliver Heaviside and Josiah Willard Gibbs), because quaternions provide superior notation. While this is undeniable for four dimensions, quaternions cannot be used with arbitrary dimensionality. Vector notation had largely replaced the "space-time" quaternions in science and engineering by the mid-20th century.

Today, the quaternions are used in computer graphics, control theory, signal processing and orbital mechanics, mainly for representing rotations/orientations. For example, it is common for spacecraft attitude-control systems to be commanded in terms of quaternions, which are also used to telemeter their current attitude. The rationale is that combining quaternion transformations is more numerically stable than combining many matrix transformations. In control and modeling applications, quaternions do not have a computational singularity that can occur for quarter-turn rotations (90 degrees) that are achievable by many Air, Sea and Space vehicles. In pure Mathematics, quaternions show up significantly as one of the four finite-dimensional normed division algebras over the real numbers, with applications throughout algebra and geometry.

Achievements

- He was twice awarded the Cunningham Medal of the Royal Irish Academy.
- The first award, in 1834, was for his work on conical refraction, for which he also received the Royal Medal of the Royal Society the following year.
- He was to win it again in 1848.
- In 1835, being secretary to the meeting of the British Association which was held that year in Dublin, he was knighted by the lord-lieutenant.
- He was successful in his election in 1837 to the president's chair in the Royal Irish Academy, and the rare distinction of being made a corresponding member of the Saint Petersburg Academy of Sciences.
- In 1864, the newly established United States National Academy of Sciences elected its first Foreign Associates, and decided to put Hamilton's name on top of their list.

Death

Irish commemorative coin celebrating the 200th Anniversary of his birth

Hamilton retained his faculties unimpaired to the very last, and steadily continued the task of finishing the Elements of Quaternions which had occupied the last six years of his life. He died on 2 September 1865, following a severe attack of gout precipitated by excessive drinking and overeating. He is buried in Mount Jerome Cemetery in Dublin. He had married Helen Bayly and had several children.

Hamilton is recognized as one of Ireland's leading scientists and as Ireland becomes more aware of its scientific heritage, he is increasingly celebrated. The Hamilton Institute is an applied Mathematics research institute at NUI Maynooth and the Royal Irish Academy holds an annual public Hamilton lecture at which Murray Gell-Mann, Frank Wilczek, Andrew Wiles and Timothy Gowers have all spoken. The year 2005 was the 200th anniversary of Hamilton's birth and the Irish government designated that the Hamilton Year, celebrating Irish science. Trinity College Dublin marked the year by launching the Hamilton Mathematics Institute. A commemorative coin was issued by the Central Bank of Ireland in his honour.

Chapter 30

Carl Gustav Jacob Jacobi

First Jewish mathematician to become a professor at a German university.

Born : **December 10, 1804, Potsdam**
Died : **September 2, 1865, Berlin**
Nationality : **German**

Early Life

Jacobi was born of Ashkenazi Jewish parentage in Potsdam on December 10, 1804. He was the second of the four children of banker Simon Jacobi. His elder brother Moritz von Jacobi would also become known later as an engineer and physicist. Jacobi was initially tutored by his uncle, Lehman who taught him the basics of classical language and elements of Mathematics. Only at the age of twelve did Jacobi enter Potsdam Gymnasium for a formal schooling. As a result of the good education received from his uncle, as well as his own remarkable abilities, after less than half a year Jacobi was moved to the senior year despite his young age. However, as the University was not accepting students younger than 16 years old, he had to remain in the senior class until 1821. He used this time to advance his knowledge, showing interest in all subjects, including Latin and Greek, philology, history and Mathematics. During this period he also made the first attempts at research trying to solve the quintic equation by radicals.

In 1821, he enrolled at the Berlin University. In the initial year, he juggled between his love for philology and Mathematics. He caught the attention of his professors with his knowledge and understanding of the subjects.

Since the university taught low level of Mathematics, he continued with his private study of the more advanced works of Euler, Lagrange and Laplace. In 1823, he finally gave up on philology to pursue his first love, Mathematics.

In 1823, he was qualified as a secondary school teacher for the subjects, Mathematics, Greek and Latin. Subsequently, he was offered a position at the Joachimsthal Gymnasium in Berlin but he declined it to pursue a university position.

In 1825, he secured a doctorate degree in philosophy. His thesis provided an analytical discussion on the theory of fractions. Same year, he gave up on his Jewish roots and converted to Christianity so as to qualify for a university position.

Qualifying for teaching University classes, the 21-year-old Jacobi lectured in 1825/26 on the theory of curves and surfaces at the University of Berlin.

In 1827 he became a professor and in 1829, a tenured professor of Mathematics at Konigsberg University, and held the chair until 1842.

Jacobi suffered a breakdown from overwork in 1843. He then visited Italy for a few months to regain his health. On his return he moved to Berlin, where he lived as a royal pensioner until his death. During the Revolution of 1848 Jacobi was politically involved and unsuccessfully presented his parliamentary candidature on behalf of a Liberal club. This led, after the suppression of the revolution, to his royal grant being cut off – but his fame and reputation were such that it was soon resumed. In 1836, he had been elected a foreign member of the Royal Swedish Academy of Sciences.

Jacobi died in 1851 from a smallpox infection. His grave is preserved at a cemetery in the Kreuzberg section of Berlin, the Fried of I der Dreifaltigkeits-Kirchengemeinde (61 Baruther Street). His grave is close to that of Johann Encke, the astronomer. The crater Jacobi on the Moon is named after him.

Scientific Contributions

- One of Jacobi's greatest accomplishments was his theory of elliptic functions and their relation to the elliptic theta function. This was developed in his great treatise *Fundamental nova theoriae functionum ellipticarum* (1829), and in later papers in Crelle's Journal. Theta functions are of great importance in mathematical Physics because of their role in the inverse problem for periodic and quasi-periodic flows. The equations of motion are integrable in terms of Jacobi's elliptic functions in the well-known cases of the pendulum, the Euler top, the symmetric Lagrange top in a gravitational field and the Kepler problem (planetary motion in a central gravitational field).
- He also made fundamental contributions in the study of differential equations and to rational mechanics, notably the Hamilton–Jacobi theory.

- It was in algebraic development that Jacobi's peculiar power mainly lay, and he made important contributions of this kind to many areas of Mathematics, as shown by his long list of papers in Crelle's Journal and elsewhere from 1826 onwards. One of his maxims was: 'Invert, always invert' ('man muss immer umkehren'), expressing his belief that the solution of many hard problems can be clarified by re-expressing them in inverse form.
- In his 1835 paper, Jacobi proved the following basic result classifying periodic (including elliptic) functions: *If a univariate single-valued function is multiply periodic, then such a function cannot have more than two periods, and the ratio of the periods cannot be a real number.* He discovered many of the fundamental properties of theta functions, including the functional equation and the Jacobi triple product formula, as well as many other results on q-series and hyper geometric series.
- The solution of the **Jacobi inversion problem** for the hyper elliptic Abel map by Weierstrass in 1854 required the introduction of the hyper elliptic theta function and later the general Riemann theta function for algebraic curves of arbitrary genus. The complex torus associated to a genus g algebraic curve, obtained by quotienting $\mathbf{C}^g$ by the lattice of periods is referred to as the Jacobian variety. This method of inversion, and its subsequent extension by Weierstrass and Riemann to arbitrary algebraic curves, may be seen as a higher genus generalization of the relation between elliptic integrals and the Jacobi, or Weierstrass elliptic functions.
- Jacobi was the first to apply elliptic functions to number theory, for example proving of Fermat's two-square theorem and Lagrange's four-square theorem, and similar results for 6 and 8 squares.
- His other work in number theory continued the work of C. F. Gauss: new proofs of quadratic reciprocity and introduction of the Jacobi symbol; contributions to higher reciprocity laws, investigations of continued fractions, and the invention of Jacobi sums.
- He was also one of the early founders of the theory of determinants; in particular, he invented the Jacobian determinant formed from the n^2 differential coefficients of n given functions of n independent variables, and which has played an important part in many analytical investigations. In 1841 he reintroduced the partial derivative ∂ notation of Legendre, which was to become standard.
- Students of vector fields and Lie theory often encounter the Jacobi identity, the analog of associativity for the Lie bracket operation.
- Planetary theory and other particular dynamical problems likewise occupied his attention from time to time. While contributing to celestial mechanics, he introduced the Jacobi integral (1836) for a sidereal coordinate system. His theory of the *last multiplier* is treated in *Vorlesungen über Dynamik*, edited by Alfred Clebsch (1866).

- He left many manuscripts, portions of which have been published at intervals in Crelle's Journal. His other works include *Commentatio de transformation integrals duplicis indefinite in form am simpliciorem* (1832), *Canon arithmetic us* (1839), and *Opuscula mathematical* (1846–1857). His *Gesammelte* (1881–1891) was published by the Berlin Academy.

Awards and Achievements

In 1836, he was elected as a foreign member of the Royal Swedish Academy of Sciences.

Personal Life and Legacy

Carl Gustav Jacob Jacobi died due to small pox on February 18, 1851. He was buried in Berlin.

Due to his eminence as a great German mathematician, his grave has been preserved at a cemetery in the Kreuzberg section of Berlin.

A crater on the moon has been named after him. Several theorems, equations, algorithms, sums, polynomials, symbols and functions in Mathematics bear his name thereby acknowledging his vast contribution to the field.

Bala Amutha P

1st Year B.Ed Student

Chapter 31

Narendra Krishna Karmarker

All India Rank 1 in the Joint Entrance Examination conducted by the prestigious IITs (IIT-JEE)

Born : 1957, Gwalior
Nationality : Indian

Narendra Krishna Karmarkar (born 1957) is an Indian mathematician, who developed Karmarkar's algorithm. He is listed as an ISI highly cited researcher.

Life History

Narendra Karmarkar was born in a Maharashtrian family in Gwalior. After securing an All India Rank 1 in the Joint Entrance Examination conducted by the prestigious IITs (IIT-JEE), he took admission in the Indian Institute of Technology Bombay. Karmarkar received his B.Tech in Electrical Engineering from IIT Bombay in 1978, M.S. from the California Institute of Technology and Ph.D. in Computer Science from the University of California, Berkeley in 1983 under the supervision of Richard M. Karp.

He invented a polynomial algorithm for linear programming also known as the interior point method. The algorithm is a cornerstone in the field of Linear Programming. He published his famous result in 1984 while he was working for Bell Laboratories in New Jersey. Karmarkar was a professor at the Tata

Institute of Fundamental Research in Mumbai. He is currently working on a new architecture for supercomputing.

Awards

- Paris Kanellakis Award, 2000 given by The Association for Computing Machinery for "specific theoretical accomplishments that have had a significant and demonstrable effect on the practice of computing".
- Srinivasa Ramanujan Birth Centenary Award for 1999, presented by the Prime Minister of India.
- Distinguished Alumnus Award, Indian Institute of Technology, Bombay, 1996
- Distinguished Alumnus Award, Computer Science and Engineering, University of California, Berkeley (1993)
- Fulkerson Prize in Discrete Mathematics given jointly by the American Mathematical Society and Mathematical Programming Society (1988)
- Fellow of Bell Laboratories (1987–)
- Texas Instruments Founders' Prize (1986)
- Marconi International Young Scientist Award (1985)
- American Academy of Achievement award, presented by former U.S. president (1985)
- Frederick W. Lanchester Prize of the Operations Research Society of America for the Best Published Contributions to Operations Research (1984)
- President of India GOLD medal, I.I.T. Bombay (1978)

Contribution to Mathematics

Karmarkar's algorithm solves linear programming problems in polynomial time. These problems are represented by "n" variables and "m" constraints. The previous method of solving these problems consisted of problem representation by an "x" sided solid with "y" vertices, where the solution was approached by traversing from vertex to vertex. Karmarkar's novel method approaches the solution by cutting through the above solid in its traversal. Consequently, complex optimization problems are solved much faster using the Karmarkar algorithm. A practical example of this efficiency is the solution to a complex problem in communications network optimization where the solution time was reduced from weeks to days. His algorithm thus enables faster business and policy decisions. Karmarkar's algorithm has stimulated the development of several other interior point methods, some of which are used in current codes for solving linear programs.

Galois geometry

After working on the Interior Point Method, Karmarkar worked on a new architecture for supercomputing, based on concepts from finite geometry, especially projective geometry over finite fields.

Current investigations

Currently, he is synthesizing these concepts with some new ideas he calls sculpturing free space (a non-linear analogue of what has popularly been described as folding the perfect corner).This approach allows him to extend this work to the physical design of machines. He is now publishing updates on his recent work including an extended abstract.This new paradigm was presented at IVNC, Poland on 16 July 2008 and at MIT on 25 July 2008. Some of his recent work is published at ieeexplore.He also delivered a lecture on his recent work at IIT Bombay in September 2013.

Chapter 32

Christian Felix Klein

Klein bore the title of Geheimrat.

Born : April 25, 1849, Dusseldorf
Died : June 22, 1925, Gottingen
Nationality : German

Christian Felix Klein was a German mathematician and Mathematics educator, known for his work in group theory, complex analysis, non-Euclidean geometry, and on the connections between geometry and group theory. His 1872 Erlangen Program, classifying geometries by their underlying symmetry groups, was a hugely influential synthesis of much of the Mathematics of the day.

Life History

Felix Klein was born on 25th April 1849 in Dusseldorf, to Prussian parents; his father, Caspar Klein (1809–1889), was a Prussian government official's secretary stationed in the Rhine Province. Klein's mother was Sophie Elise Klein (1819–1890, née Kayser). He attended the Gymnasium in Dusseldorf, then studied Mathematics and Physics at the University of Bonn, 1865–1866, intending to become a physicist. At that time, Julius Plucker held Bonn's chair of Mathematics and experimental Physics, but by the time Klein became his assistant, in 1866, Plucker's interest was geometry. Klein received his doctorate, supervised by Plucker, from the University of Bonn in 1868.

Plucker died in 1868, leaving his book on the foundations of line geometry incomplete. Klein was the obvious person to complete the second part of Plucker's *Neue Geometrie des Raumes*, and thus became acquainted with Alfred Clebsch, who had moved to Gottingen in 1868. Klein visited Clebsch the following year, along with visits to Berlin and Paris. In July 1870, at the outbreak of the Franco-Prussian War, he was in Paris and had to leave the country. For a short time, he served as a medical orderly in the Prussian army before being appointed lecturer at Gottingen in early 1871.

Erlangen appointed Klein professor in 1872, when he was only 23. He was pleased to be offered a chair at Munich's Technische Hochschule in 1875. There he and Alexander von Brill taught advanced courses to many excellent students, including, Adolf Hurwitz,Walther von Dyck, Karl Rohn, Carl Runge, Max Planck, Luigi Bianchi, and Gregorio Ricci-Curbastro.

After five years at the Technische Hochschule, Klein was appointed to a chair of geometry at Leipzig. There his colleagues included Walther von Dyck, Rohn,Eduard Study and Friedrich Engel. Klein's years at Leipzig, 1880 to 1886, fundamentally changed his life.

Klein accepted a chair at the University of Gottingen in 1886. From then until his 1913 retirement, he sought to re-establish Gottingen as the world's leading Mathematics research center. Yet he never managed to transfer from Leipzig to Gottingen his own role as the leader of a school of geometry. At Gottingen, he taught a variety of courses, mainly on the interface between Mathematics and Physics, such as mechanics and potential theory.

The research center Klein established at Gottingen served as a model for the best such centers throughout the world. He introduced weekly discussion meetings, and created a mathematical reading room and library. In 1895, Klein hired David Hilbert away from Konigsberg; this appointment proved fateful, because Hilbert continued Gottingen's glory until his own retirement in 1932.

Contribution to Mathematics

Klein's dissertation, on line geometry and its applications to mechanics, classified second degree line complexes using **Weierstrass's theory** of elementary divisors.

Klein's first important mathematical discoveries were made in 1870. In collaboration with Sophus Lie, he discovered the fundamental properties of the asymptotic lines on the Kummer surface. They went on to investigate W-curves, curves invariant under a group of projective transformations. It was Lie who introduced Klein to the concept of group, which was to play a major role in his later work. Klein also learned about groups from Camille Jordan.

Klein devised the bottle named after him, a one-sided closed surface which cannot be embedded in three-dimensional Euclidean space, but it may be immersed as a cylinder looped back through itself to join with its other end from the "inside". It may be embedded in Euclidean space of dimensions 4 and higher.

In the 1890s, Klein turned to mathematical Physics, a subject from which he had never strayed far, writing on the gyroscope with Arnold Sommerfeld. In 1894 he launched the idea of an encyclopedia of Mathematics including its applications, which became the Enzyklopädie der mathematischen Wissenschaften. This enterprise, which ran until 1935, provided an important standard reference of enduring value.

Erlangen program

In 1871, while at Gottingen, Klein made major discoveries in geometry. He published two papers *On the So-called Non-Euclidean Geometry* showing that Euclidean and non-Euclidean geometries could be considered metric spaces determined by a Cayley-Klein metric. This insight had the corollary that non-Euclidean geometry was consistent if and only if Euclidean geometry was, putting Euclidean and non-Euclidean geometries on the same footing, and ending all controversy surrounding non-Euclidean geometry. Cayley never accepted Klein's argument, believing it to be circular.

Klein's synthesis of geometry as the study of the properties of a space that is invariant under a given group of transformations, known as the *Erlangen Program* (1872), profoundly influenced the evolution of Mathematics. This program was set out in Klein's inaugural lecture as professor at Erlangen, although it was not the actual speech he gave on the occasion. The program proposed a unified approach to geometry that has become the accepted modern view. Klein showed how the essential properties of a given geometry could be represented by the group of transformations that preserve those properties. Thus the program's definition of geometry encompassed both Euclidean and non-Euclidean geometry.

Today the significance of Klein's contributions to geometry is more than evident, but not because those contributions are now seen as strange or wrong. On the contrary, those contributions have become so much a part of our present mathematical thinking that it is hard for us to appreciate their novelty, and the way in which they were not immediately accepted by all his contemporaries.

Complex analysis

Klein saw his work on complex analysis as his major contribution to Mathematics, specifically his work on:

- The link between certain ideas of Riemann's and invariant theory,
- Number theory and abstract algebra;
- Group theory;
- Geometry with more than 3 dimensions and differential equations, especially equations he invented, namely elliptic modular functions and automorphic functions.

Klein showed that the modular group moves the fundamental region of the complex plane so as to tessellate that plane. In 1879, he looked at the action of PSL(2,7), thought of as an image of the modular group, and obtained an explicit representation of a Riemann surface today called the Klein quartic. He showed that that surface was a curve in projective space, that its equation was $x^3y + y^3z + z^3x = 0$, and that its group of symmetries was PSL (2,7) of order 168. His *Ueber Riemann's Theorie der algebraischen Funktionen und ihre Integrale* (1882) treats complex analysis in a geometric way, connecting potential theory and conformal mappings. This work drew on notions from fluid dynamics. Klein considered equations of degree > 4, and was especially interested in using transcendental methods to solve the general equation of the fifth degree. Building on the methods of Hermite and Kronecker, he produced similar results to those of Brioschi and went on to completely solve the problem by means of the icosahedral group. This work led him to write a series of papers on elliptic modular functions.

In his 1884 book on the icosahedron, Klein set out a theory of automorphic functions, connecting algebra and geometry. However Poincare published an outline of his theory of automorphic functions in 1881, which led to a friendly rivalry between the two men. Both sought to state and prove a grand uniformization theorem that would serve as a capstone to the emerging theory. Klein succeeded in formulating such a theorem and in sketching a strategy for proving it. But while doing this work his health collapsed, as mentioned above. Klein summarized his work on automorphic and elliptic modular functions in a four volume treatise, written with Robert Fricke over a period of about 20 years.

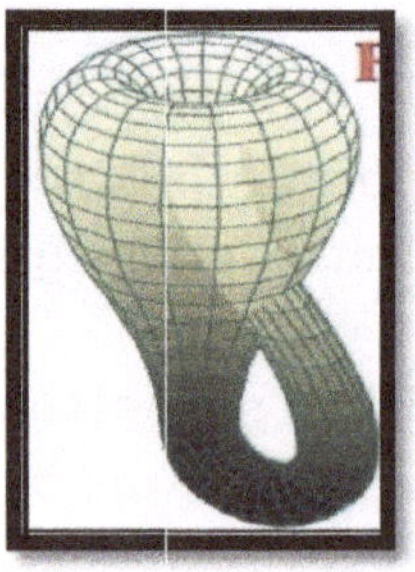

Other Researches

- He also combined his ideas on W-curves with the Norwegian Mathematician, Sophus Lie. Together they also discovered the fundamental properties of the asymptotic lines present on the Kummer surface.
- Klein published two papers on non-Euclidean geometry in the year 1871. He proved that Non-Euclidean geometry was consistent, if and only if Euclidean geometry was consistent. This discovery enabled non-Euclidean geometry to become similar to Euclidean geometry.
- His major contribution to Mathematics was his work on functional theory. In 1882, he presented a paper on the basis of treating functional theory in the geometric way by connecting conformal mappings and potential theory. Using fluid dynamics, he was able to incorporate physical ideas into his work.

- Klein worked on solving general equations of the fifth degree by using transcendental methods. He applied views of Charles Hermite and Leopold Kronecker into his methods and solved the problems with the icosahedron group. This research led him to work on elliptic modular functions.
- Klein connected geometric and algebraic results in order to develop the theory of automorphic functions, which he published in his book on icosahedron, in 1884.
- In 1900, Klein took to advocating in modernizing mathematic instructions.
- In 1905, he introduced the rudiments of calculus and the functional concepts in the secondary schools in Germany. Under Klein's direction, the German branch of the commission published several volumes for teaching Mathematics at all levels of Germany.

Publications

From the year 1874, Klein served as editor of 'Annals of Mathematics', which was one of the world's best mathematical journals. He also supervised the 'Encyclopedia of Pure and Applied Mathematics' in 1895. His papers were published in the 'Collected Mathematical Treatises' in 1923.

Achievements

- In 1908, Klein was elected as the Chairman of the 'International Commission on Mathematical Instruction' at the Rome 'International Congress of Mathematicians'.
- He was also elected as a Fellow of the 'Royal Society' in the year 1885.
- In 1912, he was awarded the 'Copley Medal of the Society'.
- He received the 'De Morgan Medal' in 1893, which was awarded by the 'London Mathematical Society'.
- He was also given the title 'Geheimrat' (Privy Councilor).

Personal Life

Klein married Anne Hegel, who was the granddaughter of the philosopher Georg Wilhelm Friedrich Hegel, in the year 1875.

Death

Klein passed away on 1925, in Gottingen.

Chapter 33

Jules Henri Poincare

It is through science that we prove, but through intuition that we discover.

Born : **April 29, 1854, Nancy**
Died : **July 17, 1912, Paris**
Nationality : **French**

Jules Henri Poincare was a French mathematician, theoretical physicist, engineer, and a philosopher of science. He is often described as a polymath, and in Mathematics as The Last Universalist by Eric Temple Bell, since he excelled in all fields of the discipline as it existed during his lifetime.

Life History

Poincare was born on 29th April 1854 in Cité Ducale neighborhood, Nancy, Meurthe-et-Moselle into an influential family. His father Leon Poincare (1828–1892) was a professor of medicine at the University of Nancy. His adored younger sister Aline married the spiritual philosopher Emile Boutroux. Another notable member of Henri's family was his cousin, Raymond Poincare, who would become the President of France, 1913 to 1920, and a fellow member of the Académie française. He was raised in the Roman Catholic faith.

Education

During his childhood he was seriously ill for a time with diphtheria and received special instruction from his mother, Eugenie Launois (1830–1897). In

1862, Henri entered the Lycée in Nancy (now renamed the Lycée Henri Poincare in his honour, along with the University of Nancy). He spent eleven years at the Lycee and during this time he proved to be one of the top students in every topic he studied. He excelled in written composition. His Mathematics teacher described him as a "monster of Mathematics" and he won first prizes in the concours général, a competition between the top pupils from all the Lycées across France. His poorest subjects were music and physical education, where he was described as "average at best".However, poor eyesight and a tendency towards absentmindedness may explain these difficulties. He graduated from the Lycée in 1871 with a bachelor's degree in letters and sciences.

During the Franco-Prussian War of 1870, he served alongside his father in the Ambulance Corps. Poincare entered the École Polytechnique in 1873 and graduated in 1875. There he studied Mathematics as a student ofCharles Hermite, continuing to excel and publishing his first paper (*Démonstration nouvelle des propriétés de l'indicatrice d'une surface*) in 1874. From November 1875 to June 1878 he studied at the École des Mines, while continuing the study of Mathematics in addition to the mining engineering syllabus, and received the degree of ordinary mining engineer in March 1879.

As a graduate of the École des Mines, he joined the Corps des Mines as an inspector for the Vesoul region in northeast France. He was on the scene of a mining disaster at Magny in August 1879 in which 18 miners died. He carried out the official investigation into the accident in a characteristically thorough and humane way.

At the same time, Poincare was preparing for his doctorate in sciences in Mathematics under the supervision of Charles Hermite. His doctoral thesis was in the field of differential equations. It was named *Sur les propriétés des fonctions définies par les équations différences*. Poincare devised a new way of studying the properties of these equations. He not only faced the question of determining the integral of such equations, but also was the first person to study their general geometric properties. He realised that they could be used to model the behaviour of multiple bodies in free motion within the solar system. Poincare graduated from the University of Paris in 1879.

The first scientific achievement

After receiving his degree, Poincare began teaching as junior lecturer in Mathematics at the University of Caen in Normandy (in December 1879). At the same time he published his first major article concerning the treatment of a class of automorphic functions.

There, in Caen, he met his future wife, Louise Poulin d'Andesi (Louise Poulain d'Andecy) and on 20 April 1881, they married. Together they had four children: Jeanne (born 1887), Yvonne (born 1889), Henriette (born 1891), and Léon (born 1893).

Poincare immediately established himself among the greatest mathematicians of Europe, attracting the attention of many prominent mathematicians. In 1881 Poincare was invited to take a teaching position at the Faculty of Sciences of the University of Paris; he accepted the invitation. During the years of 1883 to 1897, he taught mathematical analysis in École Polytechnique.

In 1881–1882, Poincare created a new branch of Mathematics: the qualitative theory of differential equations. He showed how it is possible to derive the most important information about the behavior of a family of solutions without having to solve the equation (since this may not always be possible). He successfully used this approach to problems in celestial mechanics and mathematical Physics.

Career

He never fully abandoned his mining career to Mathematics. He worked at the Ministry of Public Services as an engineer in charge of northern railway development from 1881 to 1885. He eventually became chief engineer of the Corps de Mines in 1893 and inspector general in 1910.

Beginning in 1881 and for the rest of his career, he taught at the University of Paris (the Sorbonne). He was initially appointed as the *maître de conférences d'analyse* (associate professor of analysis). Eventually, he held the chairs of Physical and Experimental Mechanics, Mathematical Physics and Theory of Probability, and Celestial Mechanics and Astronomy.

In 1887, at the young age of 32, Poincare was elected to the French Academy of Sciences. He became its president in 1906, and was elected to the Académie française in 1909.

In 1887, he won Oscar II, King of Sweden's mathematical competition for a resolution of the three-body problem concerning the free motion of multiple orbiting bodies.

In 1893, Poincare joined the French Bureau des Longitudes, which engaged him in the synchronisation of time around the world. In 1897 Poincare backed an unsuccessful proposal for the decimalisation of circular measure, and hence time and longitude. It was this post which led him to consider the question of establishing international time zones and the synchronisation of time between bodies in relative motion.

In 1899, and again more successfully in 1904, he intervened in the trials of Alfred Dreyfus. He attacked the spurious scientific claims of some of the evidence brought against Dreyfus, who was a Jewish officer in the French army charged with treason by colleagues.

In 1912, Poincare underwent surgery for a prostate problem and subsequently died from an embolism on 17 July 1912, in Paris. He was 58 years of age. He is buried in the Poincare family vault in the Cemetery of Montparnasse, Paris.

A former French Minister of Education, Claude Allègre, has recently (2004) proposed that Poincare be reburied in the Panthéon in Paris, which is reserved for French citizens only of the highest honour.

Poincare made many contributions to different fields of pure and applied Mathematics such as: cosmology celestial mechanics, fluid mechanics, optics, electricity, telegraphy, capillarity, elasticity, thermodynamics, potential theory, quantum theory, theory of relativity and physical. He was also a populariser of Mathematics and Physics and wrote several books for the lay public.

Contributions

- algebraic topology
- the theory of analytic functions of several complex variables
- the theory of abelian functions
- algebraic geometry
- Poincare was responsible for formulating one of the most famous problems in Mathematics, the Poincare conjecture, proven in 2003 by Grigori Perelman.
- Poincare recurrence theorem
- hyperbolic geometry
- number theory
- the three-body problem
- the theory of diophantine equations
- the theory of electromagnetism
- the special theory of relativity
- In an 1894 paper, he introduced the concept of the fundamental group.
- In the field of differential equations Poincare has given many results that are critical for the qualitative theory of differential equations, for example the Poincare sphere and the Poincare map.
- Poincare on "everybody's belief" in the *Normal Law of Errors* (see normal distribution for an account of that "law")
- Published an influential paper providing a novel mathematical argument in support of quantum mechanics.

The three-body problem

The problem of finding the general solution to the motion of more than two orbiting bodies in the solar system had eluded mathematicians since Newton's time. This was known originally as the three-body problem and later the *n*-body problem, where *n* is any number of more than two orbiting bodies. The *n*-body solution was considered very important and challenging at the close of the 19th century. Indeed, in 1887, in honour of his 60th birthday, Oscar II, King of Sweden, advised byGösta Mittag-Leffler, established a prize for anyone who could find the solution to the problem. The announcement was quite specific:

Given a system of arbitrarily many mass points that attract each according to Newton's law, under the assumption that no two points ever collide, try to find a representation of the coordinates of each point as a series in a variable that is some known function of time and for all of whose values the seriesconverges uniformly.

In case the problem could not be solved, any other important contribution to classical mechanics would then be considered to be prizeworthy. The prize was finally awarded to Poincare, even though he did not solve the original problem. One of the judges, the distinguished Karl Weierstrass, said, *"This work cannot indeed be considered as furnishing the complete solution of the question proposed, but that it is nevertheless of such importance that its publication will inaugurate a new era in the history of celestial mechanics."* (The first version of his contribution even contained a serious error; for details see the article by Diacu). The version finally printed contained many important ideas which led to the theory of chaos. The problem as stated originally was finally solved by Karl F. Sundman for $n = 3$ in 1912 and was generalised to the case of $n > 3$ bodies by Qiudong Wang in the 1990s.

Algebra and number theory

Poincare introduced group theory to Physics, and was the first to study the group of Lorentz transformations. He also made major contributions to the theory of discrete groups and their representations.

Topological transformation of the torus into a mug

Topology

The subject is clearly defined by Felix Klein in his "Erlangen Program" (1872): the geometry invariants of arbitrary continuous transformation, a kind of geometry. The term "topology" was introduced, as suggested by Johann Benedict Listing, instead of previously used "Analysis situs". Some important concepts were introduced by Enrico Betti and Bernhard Riemann. But the foundation of this science, for a space of any dimension, was created by Poincare. His first article on this topic appeared in 1894.

His research in geometry led to the abstract topological definition of homotopy and homology. He also first introduced the basic concepts and invariants of combinatorial topology, such as Betti numbers and the fundamental group. Poincare proved a formula relating the number of edges, vertices and faces of n-dimensional polyhedron (the Euler–Poincare theorem) and gave the first precise formulation of the intuitive notion of dimension.

Astronomy and celestial mechanics

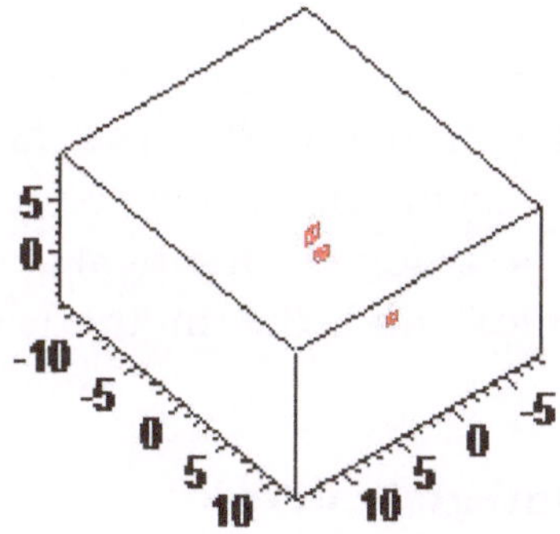

Chaotic motion in three-body problem (computer simulation)

Poincare published two now classical monographs, "New Methods of Celestial Mechanics" (1892–1899) and "Lectures on Celestial Mechanics" (1905–1910). In them, he successfully applied the results of their research to the problem of the motion of three bodies and studied in detail the behavior of solutions (frequency, stability, asymptotic, and so on). They introduced the small parameter method, fixed points, integral invariants, variational equations, the convergence of the asymptotic expansions. Generalizing a theory of Bruns (1887), Poincare showed that the three-body problem is not integrable. In other words, the general solution of the three-body problem can not be expressed in terms of algebraic and transcendental functions through unambiguous coordinates and velocities of the bodies. His works in this area were the first major achievements in celestial mechanics since Isaac Newton.

These include the idea of Poincare, who later became the base for mathematical "chaos theory" (see, in particular, the Poincare recurrence theorem) and the general theory of dynamical systems. Poincare authored important works on astronomy for the equilibrium figures gravitating rotating fluid. He introduced the important concept of bifurcation points, proved the existence of equilibrium figures of non-ellipsoid, including ring-shaped and pear-shaped figures, their stability. For this discovery, Poincare received the Gold Medal of the Royal Astronomical Society (1900).

Differential equations and mathematical Physics

After defending his doctoral thesis on the study of singular points of the system of differential equations, Poincare wrote a series of memoirs under the title "On curves defined by differential equations" (1881–1882). In these articles, he built a new branch of Mathematics, called "qualitative theory of differential equations". Poincare showed that even if the differential equation can not be solved in terms of known functions, yet from the very form of the equation, a wealth of information about the properties and behavior of the solutions can be found. In particular, Poincare investigated the nature of the trajectories of the integral curves in the plane, gave a classification of singular points (saddle, focus, center, node), introduced the concept of a limit cycle and the loop index, and showed that the number of limit cycles is always finite, except for some special cases. Poincare also developed a general theory of integral invariants and solutions of the variational equations. For the finite-difference equations, he created a new direction – the asymptotic analysis of the solutions. He applied all these achievements to study practical problems ofmathematical Physics and celestial mechanics, and the methods used were the basis of its topological works.

The singular points of the integral curves

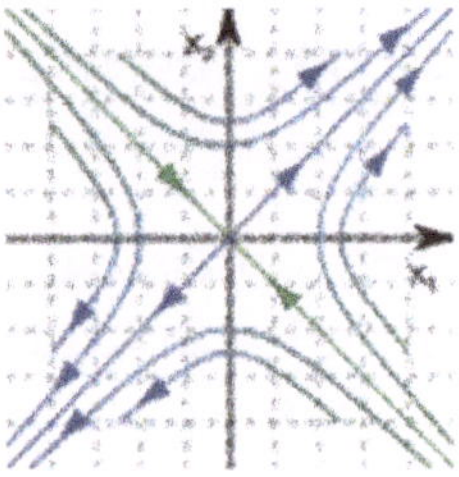

Saddle

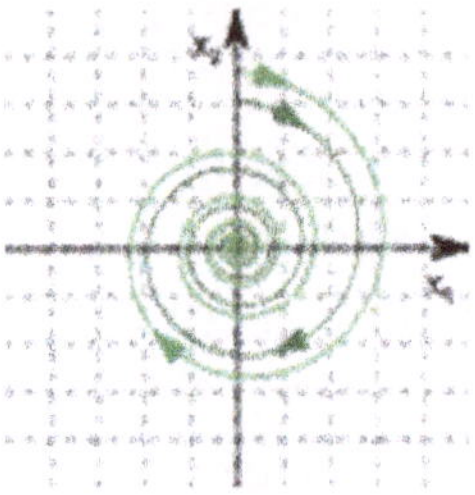

Focus

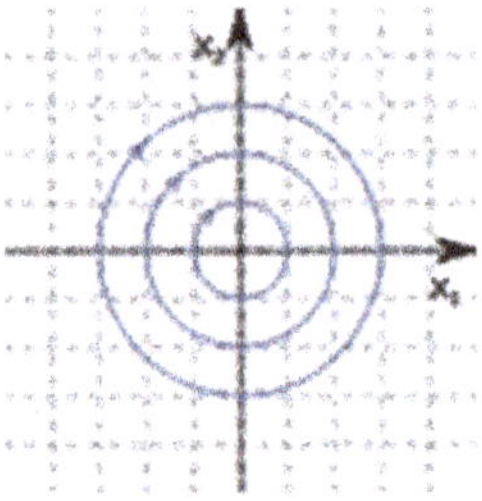

Center

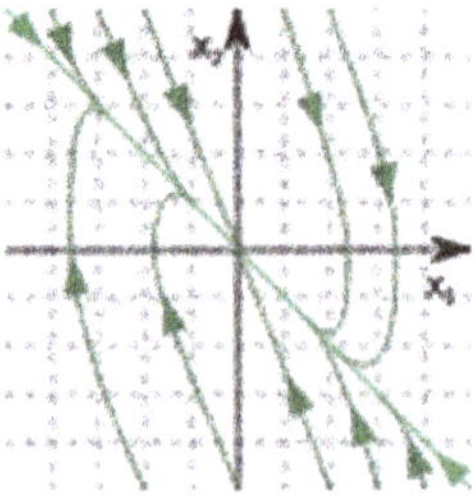

Node

Attitude towards transfinite numbers

Poincare was dismayed by Georg Cantor's theory of transfinite numbers, and referred to it as a "disease" from which Mathematics would eventually be cured. Poincare said, "There is no actual infinite; the Cantorians have forgotten this, and that is why they have fallen into contradiction."

Awards

- Oscar II, King of Sweden's mathematical competition (1887)
- Foreign member of the Royal Netherlands Academy of Arts and Sciences (1897)
- American Philosophical Society 1899
- Gold Medal of the Royal Astronomical Society of London (1900)
- Bolyai Prize in 1905
- Matteucci Medal 1905
- French Academy of Sciences 1906
- Académie française 1909
- Bruce Msedal (1911)

Avila Atreena C

1st Year B.Ed Student

Chapter 34

Brook Taylor

A lot of music is Mathematics. It's balance.

Born : **August 18, 1685, Edmonton**
Died : **November 30, 1731, London**
Nationality : **English**

Early Life

- Brook Taylor was born on 18th August 1685 in Edmonton, England. His parents John Taylor and mother Olivia Tempest had very stable financial condition.
- Taylor was home tutored before starting his studies in St. John's College, Cambridge.
- He acquired the degrees of LLB in 1709 and LLD in 1714. He was interested in Art and Music but his first love was Mathematics.
- Taylor provided the solution to the 'Kepler's Law' to Machin in 1912.
- Noticing Taylor's extraordinary expertise in the subject, he was elected as a member of the Royal Society by Machin and Keill.
- In 1714 Brook Taylor became the secretary of the Royal Society. He remained in this position for the next four years after which he resigned from the demanding position mainly due to health reasons.

Personal Life

- While highly successful in his work, his personal life suffered from many problems.
- In 1921 he fell in love with Miss Brydges who was not from a very strong and wealthy family like himself so their marriage received strong criticism by Taylor's father. The result was severed ties between father and son for the next few years.
- In 1923 his wife died in childbirth along with their first child. After receiving this strong blow Taylor went back to living with his father and their bond soon reformed.
- He married Sabetta Sawbridge in 1925. But Taylor's father only lived for four more years and died in 1729.
- The tragedy did not end here and his second wife also died while giving birth to a daughter in 1730. The child did survive this time.

Contribution

- In 1712 Taylor was elected to the Royal Society. This was on the 3rd April, and clearly it was an election based more on the expertise which Machin, Keill and others knew that Taylor had, rather than on his published results. Taylor wrote to Machin in 1712 providing a solution to a problem concerning Kepler's second law of planetary motion.
- In 1712 Taylor was appointed to the committee set up to adjudicate on whether the claim of Newton or of Leibniz to have invented the calculus was correct.
- Two books which appeared in 1715, Methodus incrementorum directa et inversa and Linear Perspective are extremely important in the history of Mathematics.

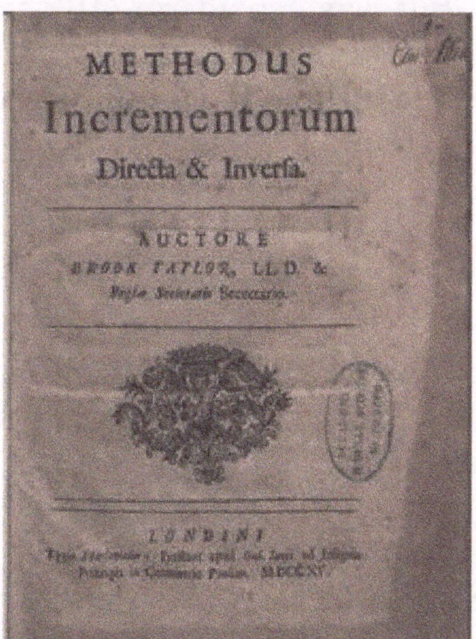

- Second editions would appear in 1717 and 1719 respectively.
- Between 1712 and 1724 Taylor published thirteen articles on topics as diverse as describing experiments in capillary action, magnetism and thermometers.
- He gave an account of an experiment to discover the law of magnetic attraction (1715) and an improved method for approximating the roots of an equation by giving a new method for computing logarithms (1717).
- Taylor added to Mathematics a new branch now called the "calculus of finite differences", invented integration by parts, and discovered the celebrated series known as Taylor's expansion.
- In fact the first mention by Taylor of a version of what is today called *Taylor's Theorem* appears in a letter which he wrote to Machin on 26 July 1712.
- The second version occurs as Corollary 2 to Proposition 7 and was thought of as a method of expanding solutions of fluxional equations in infinite series.
- The importance of Taylor's Theorem remained unrecognized until 1772 when Lagrange proclaimed it the basic principle of the differential calculus. The term *"Taylor's series"* seems to have used for the first time by Lhuilier in 1786.
- These include singular solutions to differential equations, a change of variables formula, and a way of relating the derivative of a function to the derivative of the inverse function.
- It contained is a discussion on vibrating strings, an interest which almost certainly come from Taylor's early love of music.
- Taylor, in his studies of vibrating strings was not attempting to establish equations of motion, but was considering the oscillation of a flexible string in terms of the isochrony of the pendulum. He tried to find the shape of the vibrating string and the length of the isochronous pendulum rather than to find its equations of motion.
- Taylor also devised the basic principles of perspective in *Linear Perspective* (1715).
- The second edition has a different title, being called *New principles of linear perspective.*
- At times it is very difficult for even a mathematician to understand Taylor's results. The phrase *"linear perspective"* was invented by Taylor in this work and he defined the vanishing point of a line, not parallel to the plane of the picture, as the point where a line through the eye parallel to the given line intersects the plane of the picture.
- He also defined the vanishing line to a given plane, not parallel to the plane of the picture, as the intersection of the plane through the eye parallel to the given plane.

- The main theorem in *Taylor's theory of linear perspective* is that the projection of a straight line not parallel to the plane of the picture passes through its intersection and its vanishing point.
- There is also the interesting inverse problem which is to find the position of the eye in order to see the picture from the viewpoint that the artist intended.
- Taylor was not the first to discuss this inverse problem but he did make innovative contributions to the theory of such perspective problems.
- One could certainly consider this work as laying the foundations for the theory of descriptive and projective geometry.

Death

Brook Taylor was a great mathematician who died on 30th November 1731(aged 46) London, England having given a great deal of knowledge to the world of Mathematics. He is buried in London in the churchyard in St. Anne's, Soho.

Chapter 35

Evariste Galosis

Evariste Galois has been hailed as the "father of modern algebra."

Born : October 25, 1811, Bourg-la-Reine
Died : May 31, 1832, Paris
Nationality : France

Childhood and Early Life

- Evariste Galois was born on the *25th of October, 1811* in Bourg-la-Rein, near Paris.
- Both of his parents were well educated in classical literature, religion and philosophy.

  Evariste's father, *Nicolas-Gabriel Galois*, was a Republican and headed the Bourg-la-Reine's liberal party.
- Evariste's mother, the daughter of a jurist, took care of Galois's education till he turned twelve when he entered the lycée of Louis-le-Grand in Paris in October 1823.
- Though the school was going through a great upheaval when Galois entered and about 100 students were expelled, he performed well initially and ranked first in Latin which he learnt under his mother's tutelage.

- However, he soon lost interest in studies and started taking deep interest only in Mathematics, at the age of 14.
- By February 1827 he enrolled himself for his first Mathematics class under M. Vernie.
  In 1828, Galois took the examination of the École Polytechnique, the most prestigious university of Paris, but failed to clear it.

Career

- Galois published his first Mathematics paper on continued fractions in the *Annales de mathématiques* in April, 1829.
- Around this time, he was also studying the theory of polynomial equations on which he submitted two papers, on 25 May and 1 June, to the Académie des Sciences.
- Galois made another attempt to enter the Poly technique but failed yet again.
- However, different accounts suggest different reasons behind his failure.
- After failing to secure a seat at the Poly technique, Galois took the Baccalaureate examinations in order to enter the École Normale.
- He cleared with a degree on the *29th of December 1829.*
- His Mathematics examiner reported, *"This pupil is sometimes obscure in expressing his ideas, but he is intelligent and shows a remarkable spirit of research."*
- On the other hand his literature examiner reported, *"This is the only student who has answered me poorly, he knows absolutely nothing. I was told that this student has an extraordinary capacity for Mathematics".*
- Galois sent Cauchy some of his works on the theory and then came across an article by Abel which overlapped with a part of his work.
- Cauchy suggested Galois to submit a new article on the condition that *'an equation be soluble by radicals'* in February 1830.
- The paper was then forwarded to Fourier, the secretary of the Paris Academy, to be considered for the Grand Prix of the Academy in Mathematics.
- But unfortunately, in April 1830, Fourier passed away, Galois's paper was lost and he couldn't be considered for the prize. Even amongst such disappointments Galois published three more papers that year.
- One of his papers laid the foundations for *Galois Theory.*
- The second paper was concerning *the numerical resolution of equations* and the third one was a significant work in *number theory,* where the concept of a finite field was articulated for the first time.
- Galois was affiliated to republican organisations like Republican Artillery of the National Guard and the Society of the Friends of the People and split his time between Mathematics and politics.

Contributions to Mathematics

> His work laid the foundations for ***Galois Theory & Group Theory***, two major branches of ***abstract algebra***, and ***the subfield of Galois***

- In his short lifetime, he did some phenomenal research work on Mathematics and published many of his works.
- At such a young age, Galois worked out algebraic applications of finite groups, now known as Galois groups, and laid the foundations for the solvability of algebraic equations using rational operations and extraction of roots.
- It is beyond doubt that his mathematical work helped a great deal in the transformation of the theory of algebraic equations.
- Along with Norwegian mathematician Niels Abel, he proved the impossibility of solving general quintic equation and polynomial equations of higher degree, in terms of a finite number of rational operations and root extractions.

Algebra

- While many mathematicians before Galois gave consideration to what are now known as groups, it was Galois who was the first to use the word group in a sense close to the technical sense that is understood today, making him among the founders of the branch of algebra known as group theory.
- He developed the concept that is today known as a normal subgroup. He called the decomposition of a group into its left and right cosets a proper decomposition if the left and right cosets coincide, which is what today is known as a normal subgroup.
- He also introduced the concept of a finite field (also known as a Galois field in his honour), in essentially the same form as it is understood today.

Analysis

- Galois also made some contributions to the theory of Abelian integrals and continued fractions.

Galois Theory

- Galois' most significant contribution to Mathematics by far is his development of Galois Theory.

- He realized that the algebraic solution to a polynomial equation is related to the structure of a group of permutations associated with the roots of the polynomial, the Galois group of the polynomial.
- He found that an equation could be solved in radicals if one can find a series of subgroups of its Galois group, each one normal in its successor with abelian quotient, or its Galois group is solvable.

In his last letter to Chevalier and attached manuscripts, the second of three, he made basic studies of linear groups over finite fields:

- He constructed the general linear group over a prime field, GL(ν, p) and computed its order, in studying the Galois group of the general equation of degree pν.
- He constructed the projective special linear group PSL(2,p).
- Galois constructed them as fractional linear transforms, and observed that they were simple except if p was 2 or 3.These were the second family of finite simple groups, after the alternating groups.
- He noted the exceptional fact that PSL(2,p) is simple and acts on p points if and only if p is 5, 7, or 11.

Death

Galois's died on account of a duel that occurred on the 30th May 1832. Though the reason behind the incident is not clear, there have been a great many speculations. He was discovered, hours later, by a peasant and was taken to the hospital where he passed away the next morning after speaking his final words to his brother Alfred. He died at the age of 20.

Chapter 36

Sofia Kovalevskaya

Say what you know, do what you must, come what may.

Born : **January 15, 1850, Moscow**
Died : **February 10, 1891, Stockholm, Sweden**
Nationality : **Russian**

Education and Work

Sofia Kovalevskaya was the middle child of Vasily Korvin-Krukovsky, an artillery general, and Yelizaveta Shubert, both well-educated members of the Russian nobility. Sofia was educated by tutors and governesses first at Polibino, the Krukovsky family country estate, near Pskov, then in St. Petersburg. She joined her family's social circle which included Dostoevsky (who even proposed to her elder sister Anna, but was rejected).

Sofia was attracted to Mathematics at a very young age. Her uncle Pyotr Krukovsky was not the brightest or most educated man, but he had a passion for Mathematics and told Sofia about squaring a circle and asymptotes, even before she was old enough to know what these words meant.

When Sofia was 11 years old, the walls of her nursery were temporarily (for a shortage of wallpaper) papered with pages of Ostrogradsky's lecture notes on differential and integral analysis from her father's university days. She noticed that certain things on the sheets she had heard mentioned by her uncle.

Sofia was forced to marry so that she could go abroad to enter higher education, as Russian universities in the 19th century were closed to women. Her father would not allow her to leave home to study at a university, and women in Russia could not live apart from their families without the written permission of their father or husband.

Meanwhile, Sofia and her elder sister Anna were part of an active young intelligentsia, who believed in the power of education to hasten a peaceful revolution of the tsarist social structure that would improve human nature, and to provide equal rights for women. So in 1868 the two sisters arranged fictitious marriages to radical compatriots.

Sofia married Vladimir Kovalevsky, a promising young paleontologist. This marriage caused problems for Sofia and, throughout its 15 years, it was a source of intermittent sorrow, exasperation and tension. Her concentration was broken by her frequent quarrels and misunderstandings with her husband.

In 1869 Sofia traveled to Heidelberg to study Mathematics and the natural sciences, only to discover that women could not matriculate at the university. Eventually she persuaded the university authorities to allow her to attend lectures unofficially, provided that she obtains the permission of each of her lecturers.
Sofia studied there successfully for three semesters and, according to the memoirs of a fellow student, immediately attracted the attention of her teachers with her mathematical ability. Professor Königsberger, the eminent chemist Kirchhoff, and all of the other professors were ecstatic over their gifted student and spoke about her as an extraordinary phenomenon.

In 1871 Kovalevskaya moved to Berlin to study with Weierstrass, Königsberger's teacher. Despite the efforts of Weierstrass and his colleagues, she was not permitted to attend courses at the university. Ironically this actually helped her since over the next four years Weierstrass tutored her privately.

By the spring of 1874, Kovalevskaya had completed three papers. Weierstrass deemed each of these worthy of a doctorate. The three papers were on partial differential equations, Abelian integrals and Saturn's Rings. The first of these is a remarkable contribution which was published in Crelle's Journal in 1875. The paper on the reduction of Abelian integrals to simpler elliptic integrals is of less importance, but it consisted of a skilled series of manipulations which showed her complete command of Weierstrass's theory.

In 1874 Kovalevskaya was granted her doctorate, summa cum laude, from Gottingen University in Germany. Despite this doctorate and letters of strong recommendation from Weierstrass, Kovalevskaya was unable to obtain an academic position. This was for a combination of reasons, but her gender was the major handicap.

In 1880 Sofia was persuaded to deliver a paper on some aspect of her work at the Congress of Natural Scientists held in St. Petersburg. In one quick stroke she caught the interest of the conservative Russian scientific establishment and reestablished herself as a mathematician to be taken seriously.

Gösta Mittag-Leffler, an ex-student of Weierstrass, was so impressed that for three years he was trying to find a professorship for her in Sweden. In the meantime Sofia moved to Paris and established professional contacts with the most distinguished mathematicians in France. Finally, in 1883, the University of Stockholm, under continued pressure from Mittag-Leffler, offered Sofia, now 'respectably' widowed, a probationary position as a private docent.

After the initial shock of her husband's death, Kovalevskaya immersed herself in mathematical work in an attempt to rid the feeling of guilt. She began to lecture in Stockholm in early 1884, was appointed to an extraordinary five year professorship in June of that year, and in June 1889 became the first woman since the physicist Laura Bassi and Maria Gaetana Agnesi to hold a chair at a European university.

During Kovalevskaya's years at Stockholm, she carried out what many consider her most important research. She lectured on the latest topics in analysis and became an editor of the new journal Acta Mathematica. She took over the task of liaison with the mathematicians of Paris and Berlin and took part in the organisation of international conferences. Her status brought her attention from society, and she began again to write reminiscences and dramas that she had enjoyed doing when young.

Kovalevskaya's further research on this subject won a prize from the Swedish Academy of Sciences in 1889, and in the same year, on the initiative of a famous Russian mathematician Pafnutiy Chebyshev, Kovalevskaya was elected a corresponding member of the Imperial Academy of Sciences. Although the Czarist government had repeatedly refused her a university position in Russia, the rules at the Imperial Academy were changed to allow the election of a woman.

Fascination with Maths

It was under the family's tutor Josef Malevich that Sofia undertook her first proper study of Mathematics, and she says that it was as his pupil that she began to feel an attraction for Mathematics so intense that she started to neglect other studies.

Sofia's father decided to put a stop to her math lessons, but she borrowed a copy of Bourdeu's Algebra, which she read at night when the rest of the household was asleep. A year later a neighbour, Professor Tyrtov, presented her family with a Physics textbook which he had written, and Sofia attempted to read it.

She did not understand the trigonometric formulae and attempted to explain them herself. Tyrtov realised that in her working with the concept of sine, she had used the same method by which it had developed historically. Tyrtov argued with Sofia's father that she should be encouraged to study Mathematics further but it was several years later that he permitted Sofia to take private lessons.

Later years and death

Kovalevskaya's last published work was a short article *Sur un théorème de M. Bruns*('On the Theorem by M. Bruns') in which she gave a new, simpler proof of Bruns's theorem on a property of the potential function of a homogeneous body. In early 1891, at the height of her mathematical powers and reputation, Kovalevskaya suddenly died of influenza complicated by pneumonia. Sofia was buried in Stockholm with a white lily wreath dangling from her gravestone from her most cherished friend, Weierstrass.

The Cauchy-Kaovalevskaya Theorem

"On the Theory of Partial Differential Equations," the first paper presented by Sofia for her Ph.D., received the high honour of being published in Crelle's journal in 1875. This was a very high honour for Kovalevsky because at the time she was still considered a novice mathematician. In this paper, Kovalevsky examined the existence theorem for partial differentiation originally posed by Cauchy. Kovalevsky generalized Cauchy's work on partial differential equations to systems of the k order using k-order derivatives. This combined theorem on partial differential equations is widely known as "The Cauchy-Kovalevsky Theorem". Sofia Kovalevsky's theorem on partial differential equations considerably simplified the original work of Cauchy to systems of nonlinear partial differential equations.

Using this theorem allows one to solve for the unknown entire function (U) given its partial k order derivative ($D_x^k U$) and given it satisfies the conditions. An entire function is a function that exists for all complex numbers of the complex number plane $\mathbb{C}^{n+1}$ thus making it holomorphic. This theorem assumes that the derivatives $f^0,\ldots, f^k$ (where k≥1) are entire functions on $\mathbb{C}^n$, n≥1, that h is an entire function on $\mathbb{C}^{n+1}$, and that U is an unknown entire function also on $\mathbb{C}^{n+1}$. Given the kth order partial derivative equation of U where it satisfies all the conditions (later shown graphically), the Cauchy-Kovalevsky Theorem allows us to determine the solution for U.

Equation for the kth order partial derivative $(D_x^k U)$:

$$D_x^k U = ax + by + \sum_{0 \le i < k, i+|j| \le k} c_{i,j} D_x^i D_y^j U + h(x, y)$$

This is graduate level math but we can understand the conditions of the theorem and can apply it to the real number plane as is done below. In the following application, the third order partial derivative of U was used (her theorem was for k order).

Applying the Cauchy-Kovalevskaya Theorem to Real Number

Let U be an unknown function that exists on $\mathbb{R}^2$. Suppose its partial derivatives f^0, f^1, and f^2 (zero, first and second order) all exist on $\mathbb{R}$ and h exists on $\mathbb{R}^2$. Then the third order partial derivative equation should be:

$$D_x^3 U = ax + by + \sum_{0 \le i < 3, i+|j| \le 3} c_{i,j} D_x^i D_y^j U + h(x, y)$$

For this example, in the notation of the original theorem, the partial differential equation is third order with respect to x and would be written as follows.

$$D_x^3 U = x - 6y + D_y^3 U + D_x^2 D_y^1 U + 3D_x D_y U - D_y U + h(x, y)$$

The initial condition functions $f^0(y)$, $f^1(y)$, and $f^2(y)$ are graphed below. Observing the graph at the line $x = 0$ allows us to determine if they all exist for all real numbers which in turn allows us to verify if a solution for U(x, y) exists.

First Derivative

$$\frac{\partial}{\partial x}\left((1 + y + y^2)x + \frac{1}{2}x^2 e^y + x^3 \sin(y)^2\right)$$

$$= 1 + y + y^2 + xe^y + 3x^2 \sin(y)^2$$

If you plug in x=0, this function represents $f^1(y)$.

$$f^1(y) = 1 + y + y^2$$

Maple Command:

$implicitplot3d\left(z = 1 + y + y^2 + x\,\exp(y) + 3\,x^2 \sin(y)^2, x = -0.05 \,..0.05, y = -5\,..3, z = 0\,..25, axes = boxed\right)$

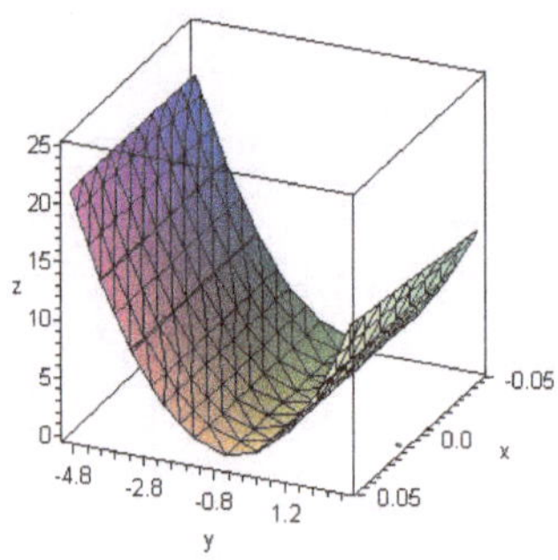

Looking at the path of the function when x=0 allows us to determine that it exists for the whole real number plane.

$$\left.\frac{dU}{dx}\right|_{x=0} = f^1(y);\ y \in \mathbb{R}$$

Second Derivative

$$\frac{\partial}{\partial x}\left(\frac{\partial}{\partial x}\left((1 + y + y^2)x + \frac{1}{2}x^2 e^y + x^3 \sin(y)^2\right)\right)$$

$$= e^y + 6x \sin(y)^2$$

If you plug in x=0, this function represents $f^1(y)$.

$$f^1(y) = e^y$$

Maple Command: implicitplot 3d

$implicitplot3d\left(z = \exp(y) + 6\,x \sin(y)^2, x = -0.04..0.04, y = -5..1, z = -0.2..3, axes = boxed\right)$

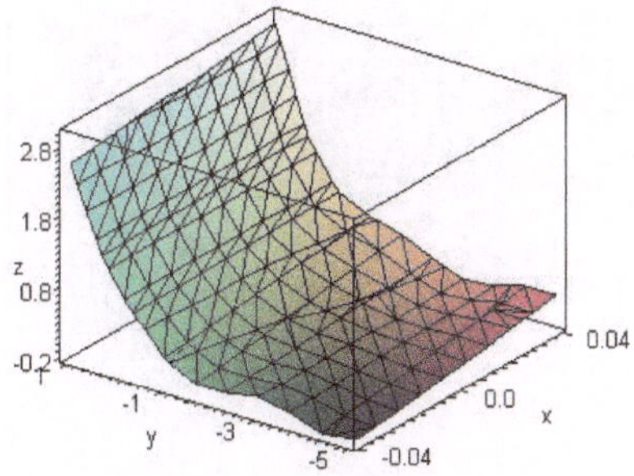

Looking at the path of the function $f^2(y)$ when x=0 allows us to determine that it exists for the whole real number plane.

$$\left.\frac{dU}{dx^2}\right|_{x=0} = f^2(y);\; y\epsilon\mathbb{R}\,,$$

Entire Function (U)

$$U = (1 + y + y^2)x + \frac{1}{2}x^2e^y + x^3\,sin(y)^2$$

If you plug in x=0, the function flattens out and is continuous.

Maple Command:

$implicitplot3d\left(z = \left(1 + y + y^2\right) x + \frac{1}{2}\, x^2 \exp(y) + x^3 \sin(y)^2, x = -0.04..0.04, y = -4..3, z = -1..1, axes = boxed\right)$

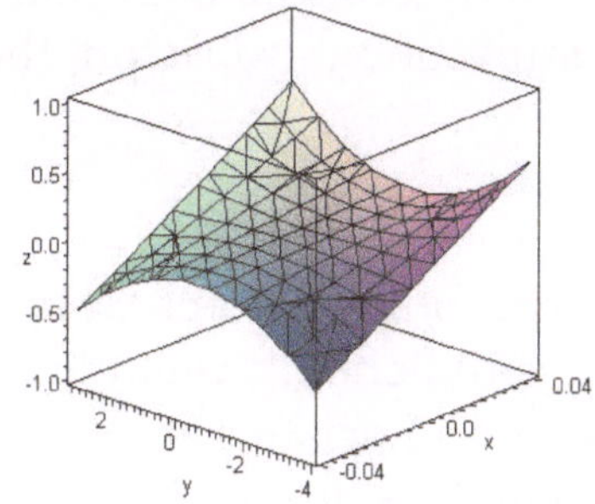

This graph of the entire function at x=0 shows that $U|_{x=0} = f^0(y);\; y\epsilon\mathbb{R}$.

Function (U) with Larger Bounds

$$U = (1 + y + y^2)x + \frac{1}{2}x^2e^y + x^3\,sin(y)^2$$

Maple Command

$implicitplot3d\left(z = \left(1 + y + y^2\right) x + \frac{1}{2}\, x^2 \exp(y) + x^3 \sin(y)^2, x = -3..3, y = -4.5..3, z = -80..100, axes = boxed\right)$

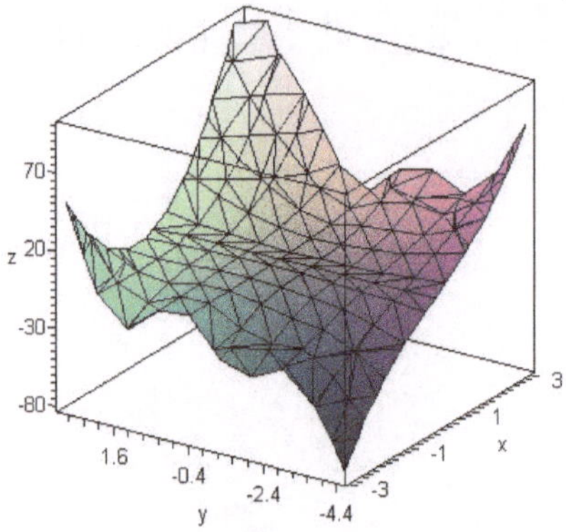

In conclusion, observing the graphs of the initial condition functions $f^0(y)$, $f^1(y)$, and $f^2(y)$ at the line x=0 permitted us to determine that they all satisfied the condition to be continuous for all real numbers.

$$U|_{x=0} = f^0(y);\ y\epsilon\mathbb{R}\ ,$$
$$\left.\frac{dU}{dx}\right|_{x=0} = f^1(y);\ y\epsilon\mathbb{R}\ ,$$
$$\left.\frac{dU}{dx^2}\right|_{x=0} = f^2(y);\ y\epsilon\mathbb{R}\ ,$$

In addition, h(x, y) exists for all (x, y) $\epsilon\,\mathbb{R}^2$ thus all the conditions for the theorem is satisfied. Therefore the Cauchy-Kovalevsky Theorem states that the given D_x^3 (the given third order partial derivative of U) has a solution for U.

Common Real Life Application

The most common application of the Cauchy-Kovalevsky Theorem is to the heat equation which is fundamental to many areas of science. The heat equation relates to the solutions of partial differential equations studied by Kovalevsky. Her work greatly contributed to subsequent work on the heat equation and is still commonly used today.

Opening Doors for Women

Mathematics was not the only subject Sofia contributed to. She had a number of literary works published, including a partly autobiographic novel *Nihilist Girl* (a story of a young aristocrat that wishes to devote her life to a cause and the effect of a changing world on the patriarchal traditions of Russia), and numerous newspaper articles on topics ranging from Sweden's peasant universities to treating hysteria by hypnosis.

An extraordinary person, Sofia Kovalevskaya was an active advocate of women's rights. It was her struggle to obtain the best education available which began to open universities' doors to women. In addition, her ground-breaking work in Mathematics made her male counterparts reconsider their archaic notions of women's inferiority to men in science, thus paving the way for future women scholars to succeed.

Hejoelin D

1st Year B.Ed Student

Chapter 37

C. R. Rao

C. R. Rao is famous for his "Theory of Estimation".

Born : **September 10, 1920 Hadagali, Kingdom of Mysore, British India**
Nationality : **Indian**
Citizenship : **United States**

Early life

C R Rao was born into a Kannada speaking Telugu family in Hadagali, Bellary, Karnataka, India. He received an M.Sc. in Mathematics from Andhra University and an M.A. in statistics from Calcutta University in 1943. He was among the first few people in the world to hold a master's degree in Statistics.

Calyampudi Radhakrishna Rao is the one of the brightest stars in the Indian sky of Statistics and Mathematics. An exceptional academician, he is a dedicated soul in the development of Statistics and his works earned him the honour of Padma Bhusan. When he was 11, he could do complicated arithmetical problems without paper and pencil. This interest should eventually get a degree in Mathematics and proceed to do research to get a doctorate degree.

Estimation Theory

- Estimation theory is a branch of statistics that deals with estimating the values of parameters based on measured/empirical data that has a random component. The parameters describe an underlying physical setting in such a way that their values affects the distribution of the measured data.
- An estimator attempts to approximate the unknown parameters using the measurements.
- In estimation theory, two approaches are generally considered.
- The probabilistic approach assumes that the measured data is random with probability distribution dependent on the parameters of interest.
- The set-membership approach assumes that the measured data vector belongs to a set which depends on the parameter vector.

The Cramer – Rao Inequality

Theorem: Let $\hat{\theta}$ be an unbiased estimator for θ. Then $var(\hat{\theta}) \geq \frac{1}{J(\theta)}$.

Proof: Using $Ev = 0$ we have:

$$E[(v - Ev)(\hat{\theta} - E\hat{\theta})] = E[v(\hat{\theta} - E\hat{\theta})]$$

$$= E[v\hat{\theta}] - E\hat{\theta}Ev$$

$$= E[v\hat{\theta}]$$

Other Contributions of C. R. Rao

- Rao defined a generalized inverse (g-inverse) of a matrix (singular or rectangular) and demonstrated its usefulness in the study of linear models and singular multivariate normal distributions.
- Rao has made some significant contributions to combinational Mathematics for use in design of experiments, the most important of which is Orthogonal arrays (OA).
- C. R. Rao started the Advanced Institute of Mathematics, Statistics and Computer Science (AIMSCS), an institute that envisages the coming together of the best minds in the fields of Mathematics, Statistics and Computer Science and working in an enabling environment. AIMSCS main purpose is to promote research and advanced studies, encourage talented students both rural and urban to pursue professional and research careers, support the application of research and training in Mathematics, Statistics and Computer Science.

Areas of Research and Contributions of C. R. Rao's are

- Estimation Theory
- Statistical inference and linear models
- Multivariate analysis
- Combinatorial design
- Orthogonal arrays Biometry

- Statistical genetics
- Generalized matrix inverses
- Functional equations

Awards and Medals

- Sardar Patel Lifetime Achievement Award (2015)
- Guy Medal in Gold (2011)
- International Mahalanobis Prize (2003)
- Srinivasa Ramanujan Medal (2003)
- National Medal of Science (2002)
- Padma Vidhushan (2001)
- Mahalanobis Centenary Gold Medal (1993)
- Wilks Memorial Award (1989)
- Padma Bhushan (1968)
- Megnadh Saha Medal (1969)
- Guy Medal in Silver (1965)
- Doctor of Science (2003)

In his honour

- The Pennsylvania State University has established C. R. and Bhargavi Rao Prize in Statistics.
- CR Rao Advanced Institute of Mathematics, Statistics and Computer Science.
- National Award in Statistics established by Ministry of Statistics and Programme Implementation (MoSPI), Government of India.
- The road from IIIT Hyderabad passing along Central University of Hyderabad crossroads to Alind Factory, Lingampally is named as "Prof. C.R. Rao Road".

Chapter 38

Harish Chandra

Developed "The Infinite Dimensional Group Representation Theory"

Born : **October 11, 1923, Kanpur, British India.**
Died : **October 16, 1983(aged 60), Princeton, New Jersey, United States**
Nationality : **Indian**
Citizenship : **United States**

Harish-Chandra was born on 11 October, 1923 in Kanpur, an industrial town near Allahabad in Uttar Pradesh, India. He became one of the leading mathematicians of the twentieth century.

Childhood

Harish Chandra Mehrotra was born in Kanpur, then known as Cawnpore in British India, to Chandrakishore Mehrotra, a civil engineer and Satyagati Seth, the daughter of a wealthy lawyer. He spent most of his childhood at his maternal grandfather's house where he received his early schooling at home from a tutor. He also learnt dancing and music. He was brilliant in his studies, but was prone to frequent illness and both these aspect of his childhood continued throughout his life. At the age of nine, Harish Chandra was enrolled in a private school and then completed his intermediate schooling from the Scindia School. He joined the

University of Allahabad to study theoretical physics in which he proved to be a brilliant student.

According to an interesting anecdote when C.V Raman was an examiner at the University, Harish Chandra solved the only question of the acoustic paper, which was on the theory of vibration of the mridangam, on the spot. He was given 100% marks by a highly impressed C.V Raman. Harish Chandra was influenced to study physics after reading the Principles of Quantum Mechanics by Paul Dirac. In 1941, he completed his B. Sc and received his Masters Degree in 1943. He then moved to the Indian Institute of Science (IISc) in Bangalore as a postgraduate research fellow under Homi Bhabha to work on problems in theoretical physics. As a research fellow, Harish Chandra published several research papers along with Bhabha, the first being 'On the Theory of Point Particles' in 1944.

He was one of the outstanding mathematicians of his generation, an algebraist and analyst, and one of those responsible for Transforming Infinite dimensional group re-presentation theory from a modest topic on the periphery of Mathematics and Physics into a major field central to contemporary Mathematics.

- The impression one has of Harish-Chandra's work as a whole is one of surprising cohesiveness and uniformity. His primary interests revolved about the representation theory of reductive groups and harmonic analysis on these groups and their related homogeneous spaces. However Harish-Chandra made several attempts to get into algebraic geometry or number theory.
- He even extended the concept of a characteristic representation of finite – dimensional representation of a case and formulated a Weyl's character formula analogue.
- Some major contributions by Harish-Chandra's work may be singled out: the explicit determination of the Plancherel measure for semisimple groups, the determination of the discrete series representations, his results on Eisenstein series and in the theory of automorphic forms, his "philosophy of cusp forms", as he called it, as a guiding principle to have a common view of certain phenomena in the representation theory of reductive groups in a rather broad sense, including not only the real Lie groups, but p-adic groups or groups over adele rings. His scientific work, being a synthesis of analysis, algebra and geometry, is still of lasting influence.

Discrete Series Representation

In Mathematics, a discrete series representation an irreducible unitary representation of a locally compact topological group G that is a subrepresentation of the left regular representation of G on $L^2(G)$. In the Plancherel measure, such representations have positive measure. The name comes from the fact that they are exactly the representations that occur discretely in the decomposition of the regular representation.

Eminent Awards

- AMS Cole Prize in Algebra (1954).
- Speaker at International Congress (1954).
- AMS Colloquium Lecturer (1969).
- Fellow of the Royal Society (1973).
- Ramanujan Medal from Indian National Science Academy (1974).
- Honourary degrees by Delhi University (1973) and Yale University (1981).
- Fellow of the National Academy of Sciences (United States) (1981).
- Indian Civilian Honour of the Padma Bhushan (1977).

Personal Life and Death

In 1952, he married Lalitha Kale the daughter of Dr. Kale a botanist when he was serving a stint at the Tata Institute. He had two daughters Premala and Devaki. In 1983, Harish Chandra was attending a conference to in honour of Armand Borel's 60th Birthday in Princeton when he had a heart attack and passed away. He had suffered three heart attacks before. He died before he could attend a similar conference in his honour and so the scheduled event became a memorial conference.

Chapter 39

Archimedes

He is known as the "Father of Geometry".

Born : **287 BC, Syracuse, Sicily Magna Graecia**
Died : **212 BC, Syracuse**
Nationality : **Greek**

His true love was pure Mathematics, and the discovery in 1906 of previously unknown works, referred to as the "Archimedes Palimpsest", has provided new insights into how he obtained his mathematical results. Today, Archimedes is widely considered to have been one of the greatest mathematicians of antiquity, if not of all time, in the august company of mathematicians such as **Newton** and **Gauss.**

Archimedes Contributions to Mathematics

- He is known as the Father of Geometry because of his many contributions to Mathematics.
- It was said that his contribution was the creation of calculus.
- These contributions are all a part of Archimedes love for creation and love for Mathematics.
- These feelings prove Archimedes to be ahead of his time by many centuries.
- Created the volume of a sphere to the volume of a cylinder ratio.

- Archimedes also is thought to be the creator of integral calculus.
- Archimedes took a circle and bisected a hexagon, until he couldn't bisect anymore, and inscribed a circle inside the hexagon.
- He noticed if he increased the number of sides he increased the accuracy of PI.
- Powers of Ten, a number which toke away the use of the Greek alphabet in the counting system.
- A formula to find the area under a curve, the amount of space that is enclosed by a curve.
- Archimedes produced formulas to calculate the areas of regular shapes, using a revolutionary method of capturing new shapes by using shapes he already understood. For example, to estimate the area of a circle, he constructed a larger polygon outside the circle and a smaller one inside it. He first enclosed the circle in a triangle, then in a square, pentagon, hexagon, etc, etc, each time approximating the area of the circle more closely. By this so-called "method of exhaustion" (or simply "Archimedes' Method").

Approximation of the area of circle by Archimedes' method of exhaustion

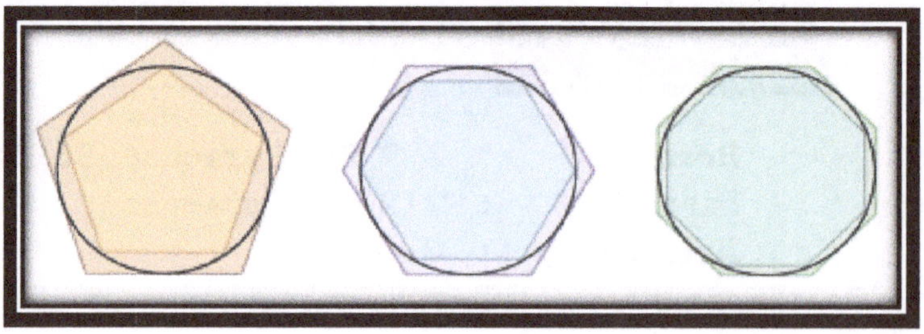

- He effectively homed in on a value for one of the most important numbers in all of Mathematics, π. His estimate was between $3\frac{1}{7}$ (approximately 3.1429) and $3\frac{10}{71}$ (approximately 3.1408), which compares well with its actual value of approximately 3.1416.
- His method for estimating π was taken to the extreme by Ludoph van Ceulen in the 16th Century, who used a polygon with an extraordinary 4,611,686,018,427,387,904 sides to arrive at a value of π correct to 35 digits. We now know that π is in fact an irrational number, whose value can never be known with complete accuracy.
- Another example of the meticulousness and precision of Archimedes' work is his calculation of the value of the square root of 3 as lying between $\frac{265}{153}$ (approximately 1.7320261) and $\frac{1351}{780}$ (approximately 1.7320512) - the actual value is approximately 1.7320508. He even calculated the number of grains of sand required to fill the universe,

using a system of counting based on the myriad (10,000) and myriad of myriads (100 million). His estimate was 8 vigintillion, or 8 x 10^{63}.

- The discovery of which Archimedes claimed to be most proud was that of the relationship between a sphere and a circumscribing cylinder of the same height and diameter. He calculated the volume of a sphere as $\frac{4}{3}\pi r^3$, and that of a cylinder of the same height and diameter as $2\pi r^3$. The surface area was $4\pi r^2$ for the sphere, and $6\pi r^2$ for the cylinder (including its two bases). Therefore, it turns out that the sphere has a volume equal to two-thirds that of the cylinder, and a surface area also equal to two-thirds that of the cylinder

Archimedes showed that the volume and surface area of a sphere are two-thirds that of its circumscribing cylinder

He also determined the area of one revolution ($0 \le \theta \le 2\pi$) of $r = a\theta$ to by

$$\text{Area} = \frac{1}{3}[\pi(a2\pi)^2]$$

- Archimedes' most sophisticated use of the method of exhaustion, which remained unsurpassed until the development of integral calculus in the 17th Century, was his proof - known as the Quadrature of the Parabola - that the area of a parabolic segment is $\frac{4}{3}$ that of a certain inscribed triangle.

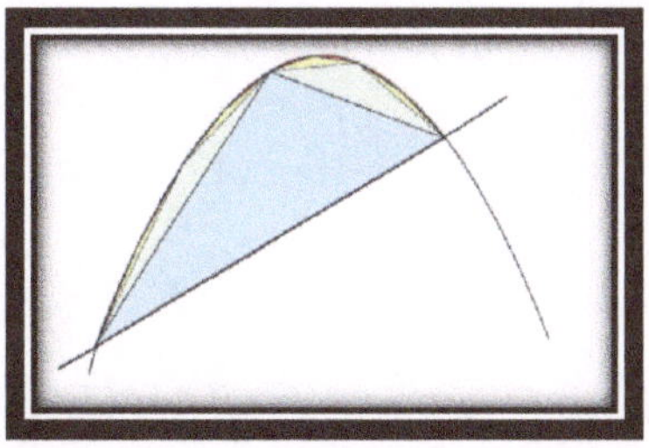

Archimedes' quadrature of the parabola using his method of exhaustion

According to legend, Archimedes was killed by a Roman soldier after the capture of the city of Syracuse. He was contemplating a mathematical diagram in the sand and enraged the soldier by refusing to go to meet the Roman general until he had finished working on the problem. His last words are supposed to have been "Do not disturb my circles".

Chellammal G

1st Year B.Ed Student

Chapter 40

Andrew Wiles

He received the Abel Prize (2016)

Born : April 11, 1953, Cambridge
Nationality : British

Place of birth

Wile is the son of Masurice Frank Wiles the Regius professor of Divinity at the University of Oxford and Patricia Wile. His father worked s the Chaplain at Ridley Hall, Cambridge, for the years 1952-55. Wile was born in Cambridge, England, in 1953, and he attended King's College School, Cambridge, and The Leys School, Cambridge.

Mathematics Career

Academic career began in the summer of 1975 under the guidance of Professor John Coates Claire College at Cambridge University, where he received his doctorate. In the period from 1977 to 1980, Wiles served as a research assistant at the college and associate professor Clare at Harvard University. Together with John Coates, he worked on the arithmetic of elliptic curves with complex multiplication methods Iwasawa theory.

One of the highlights of his career was proof of Fermat's Last Theorem in 1993 and the detection technique, which made it possible to complete the proof by his former graduate student, Robert Taylor, in 1994. Work on Fermat's theorem, he began in the summer of 1986 after Ken Ribet proved the connection of semistable elliptic curves and Fermat's theorem. The basic idea of such a connection belongs to Gerhard Fry, German mathematician, according to Andrew Wiles on solving Fermat (2000).

One of the most important events in his career was his application, he has shown great theorem Fermat in 1993 and the publication of elegant method that allows proof to be completed in 1994, professional work of the Great Fermat theorem by early summer of 1986. It can be said that the contribution of Andrew Wiles was considerable and very important for the society.

Proof of Fermat's Last Theorem

In the 1980s, most mathematicians felt that Fermat's Last Theorem, deposit its importance, was extremely hard to prove. Coates himself used the word "impossible" at one point when discussing the idea of proving it. Wiles himself was among a small minority who believed that a proof was realistic.

For more than six years, he spent all the time he had available for research on this one problem. He was very careful not to let anyone else know what he was doing, and told only his wife the full story of his research. To cover his tracks, he intermittently released carefully chosen snippets of his earlier work.

Pythagorean triples

A Pythagorean triple – named for the ancient Greek Pythagoras is a set of three integers (*a*, *b*, *c*) that satisfy a special case of Fermat's equation (n=2)

$$a^2 + b^2 = c^2$$

Examples of Pythagorean triples include (3, 4, 5) and (5, 12, 13). There are infinitely many such triples, and methods for generating such triples have been studied in many cultures, beginning with the Babylonian and later ancient Greek, Chinese, and Indian mathematicians. The traditional interest in Pythagorean triples connects with the Pythagorean triples in its converse form, it states that a triangle with sides of lengths *a*, *b*, and *c* has a right angle between the *a* and *b* legs when the numbers are a Pythagorean triple. Fermat's Last Theorem is an extension of this problem to higher powers, stating that no solution exists when the exponent 2 is replaced by any larger integer.

Diophantine equations

Fermat's equation, $x^n + y^n = z^n$ with positive integer solutions, is an example of a Diophantine equations named for the 3rd-century Alexandarian mathematician, Diophantus, who studied them and developed methods for the solution of some kinds of Diophantine equations. A typical Diophantine problem is to find two integers *x* and *y* such that their sum, and the sum of their squares, equal two given numbers *A* and *B*, respectively:

$$A = x + y; B = x^2 + y^2$$

Diophantus's major work is the Arithemetica, of which only a portion has survived. Fermat's conjecture of his Last Theorem was inspired while reading a new edition of the *Arithmetica* that was translated into Latin and published in 1621 by Claude Bachet.

Diophantine equations have been studied for thousands of years. For example, the solutions to the quadratic Diophantine equation $x^2 + y^2 = z^2$ are given by the Pythagorean triples, originally solved by the Babylonians (c. 1800 BC). Solutions to linear Diophantine equations, such as $26x + 65y = 13$, may be found using the Euclidean algorithm (c. 5th century BC). Many Diophantine equations have a form similar to the equation of Fermat's Last Theorem from the point of view of algebra, in that they have no *cross terms* mixing two letters, without sharing its particular properties. For example, it is known that there are infinitely many positive integers x, y, and z such that $x^n + y^n = z^m$ where n and m are relatively prime natural numbers.

Awards and honours

Wiles has been awarded a number of major prizes in Mathematics and science:

LMS Whitehead Prize	1988
Fellow of the Royal Society	1989
Speaker at International Congress	1994
Rolf Schock Prize	1995
Ostrowski Prize	1995
Fermat Prize	1995
Wolf Prize	1995/6
BMC plenary speaker	1996
NAS Award in Mathematics	1996
AMS Colloquium Lecturer	1996
Royal Society Royal Medal winner	1996
AMS Cole Prize in Number Theory	1997
AMS Cole Prize for number theory	1997
Clay Award	1999
LMS Honourary Member	2001
Pythagoras Award	2004
Shaw Prize	2005
Abel Prize	2016

Antony Rani Kanista A

1st Year B.Ed Student

Chapter 41

Leopold Kronecker

God made the integers, all the rest is the work of man.

Born : **December 7, 1823, Liegnitz**
Died : **December 29, 1891, Berlin**
Nationality : **German**

Kronecker's Family

Leopold Kronecker's father Isidor Kronecker was a successful business man while his mother was Johanna Prausnitzer who from a wealthy family. The families were Jewish, the religion that Kronecker kept until a year before his death when he became a convert to Christianity. Kronecker's parents employed private tutors to give him a very sound foundation to his education.

Kronecker's Interest and Education

Kronecker became a student at Berlin University in 1841. He studied under Dirichlet and Steiner. He studied not only Mathematics but also other topics such as astronomy, meteorology and chemistry. He was especially interested in philosophy studying the philosophical works of Descartes, Leibniz, Kant, Spinoza and Hegel. He went to Breslau because of his interest in Mathematics and he wanted to study again with his old school teacher Kummer who had been appointed to a chair at Breslau in 1842.

Kronecker was taught Mathematics at Liegnitz Gymnasium by Kummer. He became interested in Mathematics because of Kummer. Kummer recognised Kronecker's talent for Mathematics and encouraged him to undertake research. Despite his Jewish upbringing, Kronecker was given Evangelical religious instruction at the Gymnasium which certainly shows that his parents were open minded on religious matters.

In the year 1844-45 he worked on his doctoral thesis on algebraic number theory under Dirichlet's supervision. In 1855 Kummer came to Berlin to fill the vacancy which occurred when Dirichlet left for Gottingen. In late 1855, he took over the editorship of Crelle's Journal on Crelle's death. In 1856 Weierstrass came to Berlin, so within a year of Kronecker returning to Berlin, the remarkable team of Kummer, Borchardt, Weierstrass and Kronecker was in place in Berlin.

Of course since Kronecker did not hold a university appointment, he did not lecture at this time but was remarkably active in research publishing a large number of works in quick succession. These were on number theory, elliptic functions and algebra. He also explored the interconnections between these topics. Kummer proposed Kronecker for election to the Berlin Academy in 1860, and the proposal was seconded by Borchardt and Weierstrass. On 23rd January 1861 Kronecker was elected to the Academy and this had a surprising benefit.

Kronecker lectured at the University in the beginning in October 1862. The topics on which he lectured were very much related to his research: number theory, the theory of equations, the theory of determinants, and the theory of integrals. In his lectures

He attempted to simplify and refine existing theories and to present them from new perspectives. He did not attract great numbers of students. Only a few of his auditors were able to follow the flights of his thought, and only a few persevered until the end of the semester.

Contribution

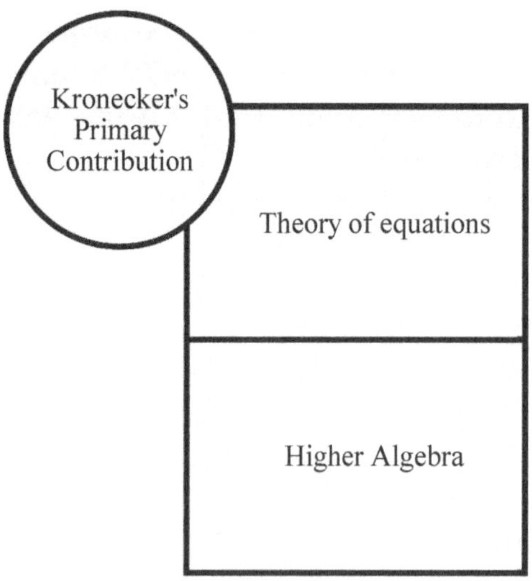

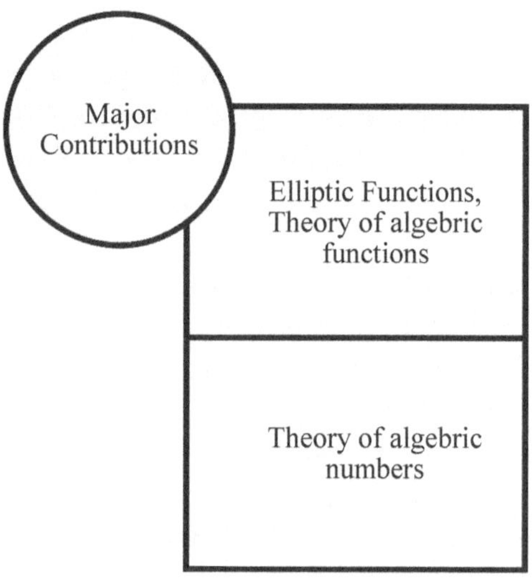

Doubting Personality

Kronecker believed that Mathematics should deal only with **finite numbers** and with a finite number of operations. He was the first to doubt the significance of non-constructive existence proofs. It appears that, from the early 1870s, Kronecker was opposed to the use of irrational numbers, upper and lower limits, and the Bolzano-Weierstrass theorem, because of their non-constructive nature. Another consequence of his philosophy of Mathematics was that to Kronecker transcendental numbers could not exist.

Editorof the journal Crell

In 1870 Heine published a paper on **trigonometric series** in Crelle's Journal, but Kronecker had tried to persuade Heine to withdraw the paper. Again in 1877 Kronecker tried to prevent publication of Cantor's work in Crelle's Journal, not because of any personal feelings against Cantor but rather because Kronecker believed that Cantor's paper was meaningless, since it proved results about mathematical objects which Kronecker believed did not exist. Kronecker was on the editorial staff of Crelle's Journal which is why he had a particularly strong influence on what was published in that journal. After Borchardt died in 1880, Kronecker took over control of Crelle's Journal as the editor and his influence on which papers would be published increased.

International Fame

The mathematical seminar in Berlin had been jointly founded in 1861 by Kummer and Weierstrass and, when Kummer retired in 1883, Kronecker became a codirector of the seminar. This increased Kronecker's influence in Berlin. Kronecker's international fame also spread, and he was honoured by being elected a foreign member of the Royal Society of London on 31 January 1884. He was also a very influential figure within German Mathematics

He established other contacts with foreign scientists in numerous travels abroad and in extending to them the hospitality of his Berlin home. For this reason his advice was often solicited in regard to filling mathematical professorships both in Germany and elsewhere.

Consistent beliefs

Although Kronecker's view of Mathematics was well known to his colleagues throughout the 1870s and 1880s, it was not until 1886 that he made these views public. Lindemann had proved that π is transcendental in 1882, and in a lecture given in 1886 Kronecker complimented Lindemann on a beautiful proof but, he claimed, one that proved nothing since transcendental numbers did not exist. So Kronecker was consistent in his arguments and his beliefs, but many mathematicians, proud of their hard earned results, felt that Kronecker was attempting to change the course of Mathematics and write their line of research out of future developments.

Kronecker explained his programme based on studying only mathematical objects which could be constructed with a finite number of operation from the integers in Über den Zahlbergriff in 1887.

Another feature of Kronecker's personality was that he tended to fall out personally with those who he disagreed with mathematically. Of course, given his belief that only finitely constructible mathematical objects existed, he was completely opposed to Cantor's developing ideas in set theory. Not only Dedekind, Heine and Cantor's Mathematics was unacceptable to this way of thinking, and Weierstrass also came to feel that Kronecker was trying to convince the next generation of mathematicians that Weierstrass's work on analysis was of no value.

Later Years

Kronecker had no official position at Berlin until Kummer retired in 1883 when he was appointed to the chair. But by 1888 Weierstrass felt that he could no longer work with Kronecker in Berlin and decided to go to Switzerland, but then, realizing that Kronecker would be in a strong position to influence the choice of his successor, he decided to remain in Berlin.

Kronecker was of very small stature and extremely self-conscious about his height. An example of how Kronecker reacted occurred in 1885 when Schwarz sent him a greeting which included the sentence:-

He who does not honour the Smaller, is not worthy of the Greater.

Here Schwarz was joking about the small man Kronecker and the large man Weierstrass. Kronecker did not see the funny side of the comment, however, and never had any further dealings with Schwarz (who was Weierstrass's student and Kummer's son-in-law). Others however displayed more tact and, for example, Helmholtz who was a professor in Berlin from 1871, managed to stay on good terms with Kronecker.

We should not think that Kronecker's views of Mathematics were totally eccentric. Although it was true that most mathematicians of his day would not agree with those views, and indeed most mathematicians today would not agree with them, they were not put aside. Kronecker's ideas were further developed by Poincare and Brouwer, who placed particular emphasis upon intuition. Intuitionism stresses that Mathematics has priority over logic, the objects of Mathematics are constructed and operated upon in the mind by the mathematician, and it is impossible to define the properties of mathematical objects simply by establishing a number of axioms.

Honours

- Kronecker was elected as a member of several academies:
    - Prussian Academy of Sciences (1861)
    - French Academy of Sciences (1868)
    - Royal Society (1884).
- The 25624 Kronecker asteroid is named after him.

Chapter 42

Jean-Victor Poncelet

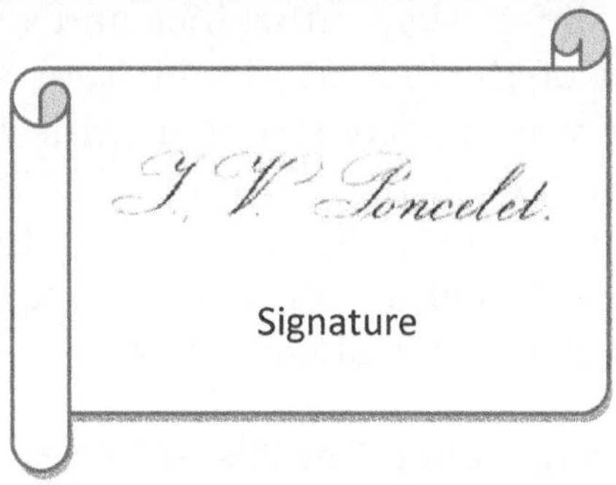

Signature

Born : July 1, 1788 Metz, Lorraine
Died : December 22, 1867, Paris
Nationality : French

History

Jean-Victor Poncelet was born in Metz, France, on July 1, 1788, the illegitimate then legitimated son of Claude Poncelet, a lawyer of the Parliament of Metz and wealthy landowner. At a young age, he was sent to live with the Olier family at Saint-Avold. He returned to Metz for his secondary education, at Lycée Fabert . After this, he attended the École Polytechnique, a prestigious school in Paris, from 1808 to 1810, though he fell behind in his studies in his third year due to poor health .After graduation, he joined the Corps of Military Engineers. He attended the École d'application in his hometown during this time, and achieved the rank oflieutenant in the French Army the same year he graduated. Poncelet took part in Napoleon's invasion of Russia in 1812.

Release and Later Employment (1822–1848)

In 1815, the year after his release, Poncelet was employed a military engineer at his hometown of Metz. In 1822, while at this position, he published *Traité des propriétés projectives des figures*. This work was the first major to discuss projective geometry since Desargues', though Gaspard Monge had written a few minor works about it previously. It is considered the founding work of modern projective geometry. Joseph Diaz Gergonne also wrote about this branch of geometry at approximately the same time, beginning in 1810. Poncelet published several papers about the subject in Gergonne's mathematical journal *Annales de Gergonne* (officially known as *Annales de mathématiques pures et appliquées*). However, Poncelet and Gergonne ultimately engaged in a bitter priority dispute over the Principle of Duality

In 1825, he became the professor of mechanics at the École d'Application in Metz, a position he held until 1835. During his tenure at this school, he improved the design ofturbines and water wheels, deriving his work from the mechanics of the Provençal mill from southern France. Although the turbine of his design was not constructed until 1838, he envisioned such a design twelve years previous to that. In 1835, his left École d'Application, and in December 1837 became a tenured professor at Sorbonne (the University of Paris), where a 'Chaire de mécanique physique et expérimentale' was specially created for him with the support of François Arago.

Commanding General at École Polytechnique (1848–1867)

In 1848, Poncelet became the Commanding General of his *alma mater*, the École Polytechnique. He held the position until 1850, when he retired. During this time, he wrote *Applications d'analyse et de géométrie*, which served as an introduction to his earlier work *Traité des propriétés projectives des figures*. It was published in two volumes in 1862 and 1864.

Contributions

Poncelet–Steiner Theorem

Poncelet discovered the following theorem in 1822: Euclidean compass and straightedge constructions can be carried out using only astraightedge if a single circle and its center is given. Swiss mathematician Jakob Steiner proved this theorem in 1833, leading to the name of the theorem. The constructions that this theorem states are possible are known as Steiner constructions.

Poncelet's Porism

In geometry, **Poncelet's porism** (sometimes referred to as **Poncelet's closure theorem**) states that whenever a polygon is inscribed in oneconic section and circumscribes another one, the polygon must be part of an infinite family of polygons that are all inscribed in and circumscribe the same two conics.

Projective Geometry

Poncelet was a French geometer who conceived of much of the theory of modern projective geometry. While confined as a Russian prisoner during the Napoleonic War of 1812, Poncelet perfected his concepts for projective geometry. Upon his release, he published them in Paris in 1822 as *Traité des propriétés projectives des figures.* The title page of the first volume of the 1865 edition is shown above. The *Traité* carried two new geometric principles that allowed for a broadening of the scope of projective geometry: the principle of duality and the principle of continuity. This work did much to advance the study of projective geometry in the nineteenth century.

He died in 1867 leaving behind works of excellent quality and in which the enormous mathematical creativity of the scientist is discovered.

Anitha K

1st Year B.Ed Student

Chapter 43

Julius Wihelm Richard Dedekind

> Numbers are the free creation of the human mind.

Born : October 6, 1831, Braunschweig, Duchy of Brunswick
Died : February 12, 1916, Braunschweig, German Empire
Nationality : German

Childhood and Early Life

Born on 6 October, 1831 in Braunschweig, Julius Wilhelm Richard Dedekind was the youngest of four children. His father Julius Levin Ulrich Dedekind was an administrator at Collegium Carolinum in his hometown. Dedekind had three elder siblings. He was educated at Martino-Catharineum school in Braunschweig and Collegium Carolinum. During school days, Dedekind did not show any particular interest in Mathematics. He studied chemistry and Physics. He was not particularly happy with vague logic of Physics, which pulled him towards Mathematics. In 1850 he entered the University of Gottingen to study under Moritz Stern. Dedekind earned his doctorate degree in 1852 for his thesis 'Über die Theorie der Eulerschen Integrale' (On the Theory of Eulerian integrals).

Career

Dedekind started his career teaching at a Polytechnic in Zürich in 1858. However, in 1862, he came back to his hometown Braunschweig to teach at Technische Hochschule, previously known as Collegium Carolinum. It was while giving lessons on calculus at Polytechnic that Dedekind came up with the concept of Schnitt or Dedekind cut, which is now considered as a 'standard definition of the real numbers'. Later, he wrote on Dedekind cuts and irrational numbers and published these works on a pamphlet namely Stetigkeit und irrationale Zahlen (Continuity and irrational numbers). Dedekind held close association with Cantor after they met in 1874. He had great admiration for Cantor's work on infinite sets. Besides his own works such as Dedekind's theorem, Dedekind edited works of Gauss, Riemann and Dirichlet. Dedekind went on to publish Dirichlet's lectures on number theory titled 'Vorlesungen über Zahlentheorie' (Lectures on Number Theory) in German in 1863.

His investigation on Dirichlet's work directed his later analysis on ideals and algebraic number fields. Dedekin also presented works on ideals. Dedekind, along with Heinrich Martin Weber, gave algebraic proof to the Riemann-Roch theorem in 1882. Later in 1888, he presented short essay entitled 'Was sind und was sollen die Zahlen' (What are numbers and what should they be?). In 1900 he published works on modular lattices in algebra. He continued working at Technische Hochschule until his retirement in 1894. He continued publishing works and teaching even after the retirement.

Personal Life and Death

Dedekind remained unmarried for his lifetime and lived with his sister Julia. He took occasional holidays in the Black Forest in Germany and in Switzerland. In one of such trip, he met Cantor- a fellow mathematician where the two got to work on set theory. He breathed his last on 12 February 1916 in Braunschweig, Germany.

Major Works

- Lectures on Number Theory
- Continuity and irrational numbers
- What are numbers and what should they be?
- Modular Lattices

Dedekind's Contributions to the Foundations of Mathematics

While many of Dedekind's contributions to Mathematics and its foundations are thus common knowledge, they are seldom discussed together. In particular, his mathematical writings are often treated separately from his foundational ones. This entry provides a broad survey of his contributions. The main focus will be on his foundational writings, but they will be related to his other mathematical work. Indeed, it will be argued that foundational concerns are at play throughout Dedekind's work, so that any attempt to distinguish sharply between his "mathematical" and "foundational" writings is artificial and misleading. Another goal of the entry is to establish the continuing relevance of his contributions to the philosophy of Mathematics whose full significance has only started to be recognized.

The Foundations of Analysis

An important part of the dichotomy, as traditionally understood, was that magnitudes and ratios of them were not thought of as numerical entities, with arithmetic operations defined on them, but in a more concrete geometric way (as lengths, areas, volumes, angles, etc. and as relations between them). More particularly, while Eudoxos' theory provides a contextual criterion for the equality of ratios, it does not provide for a definition of the ratios themselves, so that they are not conceived of as independent entities (Mueller 1981, Stein 1990, Cooke 2005). Such features do little harm with respect to empirical applications of the theory. But they lead to inner-mathematical tensions when solutions to various algebraic equations are considered (some of which could be represented numerically, but others only geometrically). This tension came increasingly to the fore in the Mathematics of the early modern period, especially after Descartes' integration of algebra and geometry. What was called for, in response, was a unified treatment of discrete and continuous quantities.

The Foundations of Arithmetic

Providing an explicit, systematic, and fruitful definition of the real numbers constitutes a major step towards completing the arithmetization of analysis. Further reflection on Dedekind's procedure (and similar ones) leads to a new question, however: What exactly is involved in it if it is thought through fully, i.e., what does his account of the real numbers rely on ultimately? As noted, Dedekind starts with the system of rational numbers; and he uses a set-theoretic procedure to construct, in a central step, the new system of cuts out of them. This suggests two sub-questions: First, how exactly are we to think about the rational numbers in this connection? Second, can anything further be said about the relevant set-theoretic procedures and the assumptions behind them?

In his published writings Dedekind does not provide a direct and full answer to our first sub-question. What suggests itself, from a contemporary point of view, is that he relied on the idea that the rational numbers can be dealt with in terms of the natural numbers together with some set-theoretic techniques.

Richard Dedekind Timeline

1831 : Dedekind was born in Braunschweig.
1850 : He got enrolled at University of Gottingen.
1858 : He began his career as a teacher at Polytechnic in Zurich.
1862 : He returned to Braunschweig to teach at Technische Hochschule.
1874 : He met Cantor, a famous mathematician.
1880 : He was elected to the Academy of Berlin and Rome.

1882 : He, along with Heinrich Martin Weber gave algebraic proof to the Riemann-Roch theorem.

1888 : He published short essay entitled 'What are numbers and what should they be?'.

1894 : He retired from teaching job at Technische Hochschule.

1900 : He presented paper on modular lattices in algebra and elected to the Académie des Sciences, Paris

1916 : He died in Braunschweig, Germany.

Chapter 44

Nikolai Ivanovich Lobachevsky

Copernicus of Geometry

Born : **December 1,1792, Nizhny Novgorod**
Died : **February 24, 1856, Kazan, Russia**
Nationality : **Russian**

Childhood and Marriage Life

Lobachevsky was born in 1792 in Nizhny Novgorod, Russia. His parents were Ivan Maksimovich Lobachevsky, a clerk in a land surveying office, and Praskovia Alexandrovna Lobachevskaya. Nikolay was one of three sons in the poor family. In 1800, Lobachevsky's father died and his mother moved to Kazan.

In 1807 Lobachevsky graduated from the Gymnasium and entered Kazan University as a free student. Kazan State University had been founded in 1804, the result of one of the many reforms of the emperor Alexander I, and it opened in the following year, only two years before Lobachevsky began his undergraduate career.

In 1832 Lobachevsky married Lady Varvara Alexeevna Moiseeva who came from a wealthy family. At the time of his marriage his wife was a young girl while Lobachevsky was forty years old.

Career

- Lobachevsky received a Master's Degree in Physics and Mathematics in 1811.
- In 1814 he was appointed to a lectureship and in 1816 he became an extraordinary professor.
- In 1822 he was appointed as a full professor.
- In 1827 Nikolay Ivanovich Lobachevsky became a rector of Kazan University and dedicated following 19 years to the university, making it flourish.
- He managed to introduce new educational standards; his reign resulted in library, astronomical and magnetic observatories, dissecting room, physical and chemical laboratories.
- For eight years Lobachevsky headed university's library and read special courses for students.
- His knowledge helped students and teachers to avoid cholera, and his enthusiasm saved books and scientific equipment from Kazan's great fire of 1842.
- He popularized science and opened free public access to the library and museums of the university.

Contribution

Fifth Postulate of Euclid

Nikollay Ivanovich Lobachevsky was the first mathematician to publish an account of non-Euclidean geometry. Since Euclid's axiomatic formulation of geometry mathematicians had been trying to prove his fifth postulate as a theorem deduced from the other four axioms. The fifth postulate states that given a line and a point not on the line, a unique line can be drawn through the point parallel to the given line. Lobachevsky did not try to prove this postulate as a theorem. Instead he studied geometry in which the fifth postulate does not necessarily hold. Lobachevsky categorised euclidean as a special case of this more general geometry.

Non-Euclidean Geometry

His major work, *Geometries* completed in 1823, was not published in its original form until 1909. On 11 February 1826, in the session of the Department of Physico-Mathematical Sciences at Kazan University, Lobachevsky requested that his work about a new geometry was heard and his paper *A concise outline of the foundations of geometry* was sent to referees. The text of this paper has not survived but the ideas were incorporated, perhaps in a modified form, in Lobachevsky's first publication on hyperbolic geometry. He published this work on non-euclidean geometry, the first account of the subject to appear in print, in 1829.

Dandelin-Gräffe method

In 1834 Lobachevsky found a method for the approximation of the roots of algebraic equations. This method of numerical solution of algebraic equations, developed independently by Gräffe to answer a prize question of the Berlin Academy, is today a particularly suitable method for using computers to solve such problems. This method is today called the Dandelin-Gräffe method since Dandelin also independently investigated it, but only in Russia does the method appear to be named after Lobachevsky who is the third independent discoverer.

Quotation

"In geometry I find certain imperfections which I hold to be the reason why this science, apart from the transition into analytics, can as yet make no advance from that state in which it has come to us from Euclid." – Nikolai Ivanovich Lobachevsky

Books

- Pangeometry
- Geometrical research on the theory of parallels

Honours

- 1858 Lobachevsk, an asteroid discovered in 1972, was named in his honour.
- The lunar crater Lobachevskiy was named in his honour.
- Lobachevsky Prize, a Mathematics award by the Kazan State University.

Later Life

A worsening sclerotic condition progressively affected his eyesight, and he was blind in his last years. Nikolay Ivanovich Lobachevsky died on February 24, 1856 in Kazan. His great mathematical achievements were not recognized in his lifetime, and he died without having any notion of the fame and importance that his work would achieve.

In recognition of his work Lobachevsky was in 1837 raised to the hereditary nobility; he designed his own heraldic device (which is reproduced on his tombstone), depicting Solomon's seal, a bee, an arrow, and a horseshoe, to symbolize wisdom, diligence, alacrity, and happiness, respectively.

Chapter 45

Gerolamo Carnado

The final line of Ars magna
I wrote it out five times,
may it last the same
number of millennia.

Born : September 24, 1501, Pavia, Duchy of milan(now Italy)
Died : September 21, 1576, Rome (now Italy)
Nationality : Italian

Early life and education

Gerolamo Cardano was an Italian Renaissance mathematician, physician, astrologer, philosopher and gambler. He wrote more than 200 works on medicine, Mathematics, physics, philosophy, religion, and music. His gambling led him to formulate elementary rules in probability, making him one of the founders of the field.

He was born in Pavia, Lombardy, the illegitimate child of Fazio Cardano, a mathematically gifted jurist and lawyer, who was a friend of Leonardo da Vinci. In his autobiography, Cardano wrote that his mother, Chiara Micheri, had taken "various abortive medicines" to terminate the pregnancy; he was "taken by violent means from my mother; I was almost dead." She was in labour for three days. Shortly before his birth, his mother had to move from Milan to Pavia to escape the Plague; her three other children died from the disease.

After a depressing childhood, with frequent illnesses and the rough upbringing by his overbearing father, in 1520, Cardano entered the University of

Pavia against his father's wish, who wanted his son to undertake studies of law, but Girolamo felt more attracted to philosophy and science. During the ongoing war between France and Spain, the authorities in Pavia were forced to close the university; Cardano resumed his studies at the University of Padua, where he graduated in medicine in 1526. His eccentric and confrontational style did not earn him many friends and he had a difficult time finding work after his studies had ended. In 1525, Cardano repeatedly applied to the College of Physicians in Milan, but was not admitted owing to his combative reputation and illegitimate birth. He suffered from impotence throughout the early part of his life, but recovered and married Lucia Banderini in 1531. Before her death in 1546, she bore him three children, Giovanni Battista (1534), Chiara (1537) and Aldo (1543).

Without a doubt, he is one of the most colorful characters in the early history of probability. Cardano's early training in Mathematics probably came from his father, who was a lawyer by trade, but evidently an accomplished mathematician as well. Later, Cadano was educated at the universities of Pavia and Padua. He received a doctoral degree in medicine in 1525.

Cardano was a man of many interests and professions, including: law, medicine, astrology, and Mathematics. Cardano published books on arithmetic and simple mensuration in 1537. The mathematical work that he is best known for, Ars Magna (The Great Art) was published in 1545. This book contains a variety of methods for solving polynomial equations, and anticipates the discovery of complex numbers.

Contribution

Today, he is best known for his achievements in algebra. Cardano was the first mathematician to make systematic use of numbers less than zero. He published the solutions to the cubic and quartic equations in his 1545 book Ars Magna. The solution to one particular case of the cubic equation ax^3+bx+c=0(in modern notation), was communicated to him by Niccolò Fontana Tartaglia in the form of a poem.The quartic was solved by Cardano's student Lodovico Ferrari. Both were acknowledged in the foreword of the book, as well as in several places within its body. In his exposition, he acknowledged the existence of what are now called imaginary numbers, although he did not understand their properties. In Opus novum de proportionibus he introduced the binomial coefficients and the binomial theorem.

Cardano was notoriously short of money and kept himself solvent by being an accomplished gambler and chess player. His book about games of chance, Liber de ludo aleae ("Book on Games of Chance"), written around 1564, but not published until 1663, contains the first systematic treatment of probability,as well as a section on effective cheating methods. He used the game of throwing dice to understand the basic concepts of probability. He demonstrated the efficacy of defining odds as the ratio of favourable to unfavourable outcomes (which implies that the probability of an event is given by the ratio of favourable outcomes to the total number of possible outcomes). He was also aware of the multiplication rule for independent events but was not certain about what values should be multiplied.Cardano invented several mechanical devices including the combination lock, the gimbal consisting of three concentric rings allowing a supported compass or gyroscope to rotate freely, and the Cardan shaft with universal joints, which allows the transmission of rotary motion at various angles and is used in vehicles to this day. He studied hypocycloids, published in de proportionibus 1570. The generating circles of these hypocycloids were later named Cardano circles or cardanic circles and were used for the construction of the first high-speed printing presses.

Cardano was described hypocycloids, the pointed plane curves generated by the trace of a fixed point on a small circle that rolls within a larger circle, and the generating circles were later named Cardano (or Cardanic) circles.

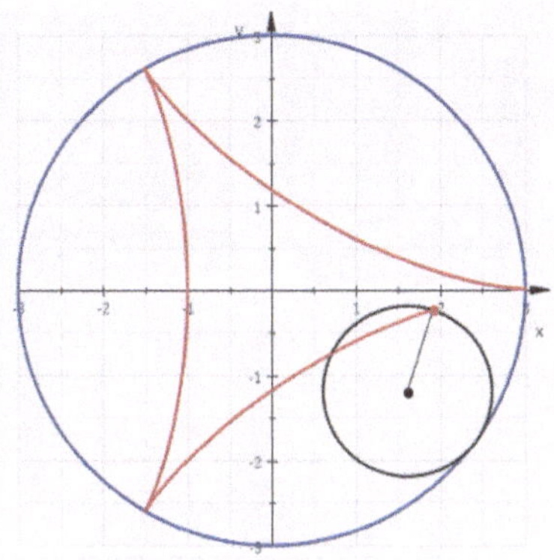

The circles used to generate hypocycloids are known as Cardano Circles

He made several contributions to hydrodynamics and held that perpetual motion is impossible, except in celestial bodies. He published two encyclopedias of natural science which contain a wide variety of inventions, facts, and occult superstitions. He also introduced the Cardan grille, a cryptographic tool, in 1550.

Someone also assigned to Cardano the credit for the invention of the so-called Cardano's Rings, also called Chinese Rings, but it is very probable that they predate Cardano.

Significantly, in the history of education of the deaf, he said that deaf people were capable of using their minds, argued for the importance of teaching them, and was one of the first to state that deaf people could learn to read and write without learning how to speak first. He was familiar with a report by Rudolph Agricola about a deaf mute who had learned to write.

Anitha G

1st Year B.Ed Student

Chapter 46

Paul Erdos

To give a mathematical lecture was "to preach".

Born : **March 26, 1913, Budapest**
Died : **September 20, 1996, Warsaw, Poland**
Nationality : **Hungarian**

Early life of Erdos

Paul Erdos was born in Budapest, Austria-Hungary, on March 26, 1913. He was the only surviving child of Anna and Lajos Erdos (formerly Englander).His two sisters, aged 3 and 5, both died of scarlet fever a few days before he was born. His fascination with Mathematics developed early by the age of four, given a person's age, he could calculate, in his head, how many seconds they had lived.

Both of Erdos's parents were high school Mathematics teachers, and Erdos received much of his early education from them. Erdos always remembered his parents with great affection. At 16, his father introduced him to two of his lifetime favourite subjects infinite series and set theory.

In 1934, at the age of 21, he was awarded a doctorate in Mathematics from the University of Budapest. On September 20, 1996, at the age of 83, he had a heart attack and died while attending a conference in Warsaw. He never married and had no children. He is buried next to his mother and father in grave 17A-6-29 at Kozma Utcai Temeto in Budapest. For his epitaph, he suggested "I've finally stopped getting dumber."

Mathematical work

- Erdos was one of the most prolific publishers of papers in mathematical history, comparable only with Leonhard Euler;
- Erdos published more papers, mostly in collaboration with other mathematicians, while Euler published more pages, mostly by himself.
- Erdos wrote around 1,525 mathematical articles in his lifetime.
- In his mathematical style, Erdos was much more of a "problem solver" than a "theory developer".
- He resolutely concentrated on particular theorems and conjectures throughout his illustrious career.
- Erdos never won the highest mathematical prize, the Fields Medal.
- He won the Wolf Prize, where his contribution is described as "for his numerous contributions to number theory, combinatorics, probability, set theory and mathematical analysis, and for personally stimulating mathematicians the world over".

Contribution to Mathematics

- The most mathematically notable of Erdos' work is the Erdos conjecture on arithmetic progressions.
- If the sum of the reciprocals of a sequence of integers diverges, then the sequence contains arithmetic progressions of arbitrary length.
- The most familiar problem with an Erdos prize is likely the Collatz conjecture, also called the 3N + 1 problem.

Erdos number

Because of his prolific output, friends created the Erdos number as a tribute. An Erdos number describes a person's degree of separation from Erdos himself, based on their collaboration with him, or with another who has their own Erdos number. Erdos alone was assigned the Erdos number of 0 (for being himself), while his immediate collaborators could claim an Erdos number of 1, their collaborators have Erdos number at most 2, and so on. The Erdos number was most likely first defined by Casper Goffman, an analyst whose own Erdos number is 2.

Signature

Erdos signed his name "Paul Erdos P.G.O.M." When he became 60, he added "L.D.", at 65 "A.D.", at 70 "L.D." and at 75 "C.D."

- P.G.O.M. represented "Poor Great Old Man"
- L.D. represented "Living Dead"
- A.D. represented "Archaeological Discovery"
- C.D. represented "Counts Dead"

Chapter 47

Emmy Noether

Greatest woman mathematician of all time.

Born : March 23, 1882, Erlangen
Died : April 14, 1935, Bryn Mawr, Pennsylvania, U.S.
Nationality : German

Emmy Noether was born on 23 March 1882, the first of four children. Her first name was "Amalie", after her mother and paternal grandmother, but she began using her middle name at a young age. She pursued none of these activities with passion, although she loved to dance.

Noether never married as she was passionate only about Mathematics. She had many friends who were colleagues and fellow mathematicians. Her closest friend was Anna Pell Wheeler, a fellow colleague and Mathematician at Bryn Mawr College. Hermann Weyl was also a dear friend of hers at Bryn Mawr College. At a point in her life, Noether was diagnosed with an illness which she spoke of only to her closest friends.

Career in Mathematics

From 1908 to 1915, she worked at the Mathematical Institute of Erlangen without pay, and piloted her researches there. Felix Klien and David Hilbert invited Noether to join the Mathematics department at the University of Gottingen in 1915. Although she was criticized by many for working at the University, she lectured students for four years under Hilbert's name. She was given the title 'Privatdozent', which permitted her to lecture in 1919, but she was still not paid. In 1922, Noether became an associate professor receiving a menial salary for her service.

Despite her brilliant works and knowledge, she was not given the status of a professor as she was a woman, a Jew and a social democrat. During the years 1928 to 1929, Noether became a guest lecturer at the University of Moscow. She taught at the University of Frankfurt in 1930. In 1932, she gave a lecture in Zurich at the International Mathematical Congress. She was a member of the Gottingen Mathematics department till 1933. When Nazis took over, she was unable to continue her profession in Germany and so, in 1933, she moved to the U.S and taught at the Bryn Mawr College in Pennsylvania as a guest professor. She was paid a full salary here and was accepted as a proper faculty member. She also taught at the Institute of Advanced Study at Princeton.

Works and Achievements

Noether published several papers while she was working at the Mathematical Institute of Erlangen. She began her research on theoretical algebra and collaborated with Algebraist, Ernst Otto Fischer, for her works. She also teamed with Felix Klein and David Hilbert to work on Einstein's general relativity theory.

Post Card written by Emmy to discuss Abstract Algebra

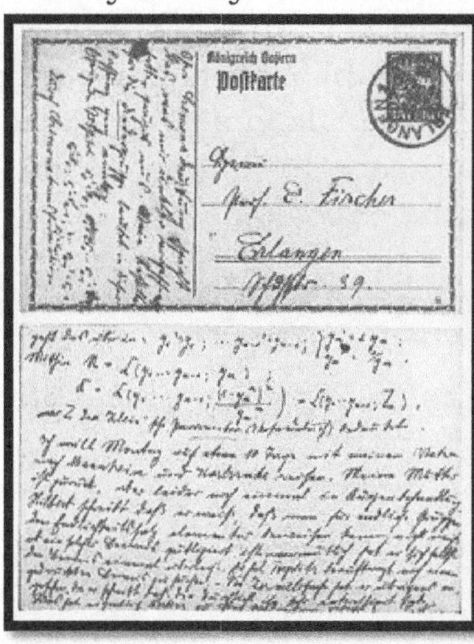

According to Hermann Weyl, Fischer was an important influence on Noether, in particular by introducing her to the work of David Hilbert. From 1913 to 1916 Noether published several papers extending and applying Hilbert's methods to mathematical objects such as fields of rational functions and the invariants of finite groups. This phase marks the beginning of her engagement with abstract algebra, the field of Mathematics to which she would make groundbreaking contributions.

Noether and Fischer shared lively enjoyment of Mathematics and would often discuss lectures long after they were over; Noether is known to have sent postcards to Fischer continuing her train of mathematical thoughts.

Mathematical Contributions

Soon after arriving at Gottingen, however, she demonstrated her capabilities by proving the theorem now known as Noether's theorem, which shows that a conservation law is associated with any differentiable symmetry of a physical system.

American physicists Leon M. Ledermanand Christopher T. Hill argue in their book Symmetry and the Beautiful Universe that Noether's theorem is "certainly one of the most important mathematical theorems ever proved in guiding the development of modern Physics, possibly on a par with the Pythagorean theorem"

Work in abstract Algebra

Although Noether's theorem had a profound effect upon Physics, among mathematicians she is best remembered for her seminal contributions to abstract algebra. In his Introduction to Noether's Collected Papers, Nathan Jacobson wrote that "The development of abstract algebra, which is one of the most distinctive innovations of twentieth century Mathematics, is largely due to her – in published papers, in lectures, and in personal influence on her contemporaries."

Noether's groundbreaking work in algebra began in 1920. In collaboration with W. Schmeidler, she then published a paper about the theory of ideals in which they defined left and right ideals in a ring. The following year she published a landmark paper called Idealtheorie in Ringbereichen, analyzing ascending chain conditions with regard to (mathematical) ideals.

Alexandrov named Emmy Noether "the greatest woman mathematician of all time".

Noether was recorded as having given at least five semester-long courses at Gottingen.

- Winter 1924/25: Gruppentheorie und hyperkomplexe Zahlen (Group Theory and Hypercomplex Numbers)
- Winter 1927/28: Hyperkomplexe Grössen und Darstellungstheorie (Hypercomplex Quantities and Representation Theory)
- Summer 1928: Nichtkommutative Algebra (Noncommutative Algebra)
- Summer 1929: Nichtkommutative Arithmetik (Noncommutative Arithmetic)
- Winter 1929/30: Algebra der hyperkomplexen Grössen (Algebra of Hypercomplex Quantities).

These courses often preceded major publications in these areas.

She worked with the topologists, Lev Pontryagin and Nikolai Chebotaryov, who later praised her contributions to the development of Galois theory.

Noether's colleagues celebrated her fiftieth birthday in 1932, in typical mathematicians' style. Helmut Hasse dedicated an article to her in the Mathematische Annalen, wherein he confirmed her suspicion that some aspects of non-commutative algebra are simpler than those of commutative algebra, by proving a non-commutative reciprocity law. This pleased her immensely. He also sent her a mathematical riddle, the "mμν-riddle of syllables", which she solved immediately; the riddle has been lost.

Contributions to Mathematics and Physics

Paul Gordan supervised Noether's doctoral dissertation on invariants of biquadratic forms. Noether's work in abstract algebra and topology was influential in Mathematics, while in Physics, Noether's theorem has far-ranging consequences for theoretical Physics anddynamic systems.

Emmy Noether's scientific production fell into three clearly distinct epochs:

1. The period of relative dependence, 1907–1919;
2. The investigations grouped around the general theory of ideals 1920–1926;
3. The study of the non-commutative algebras, their representations by linear transformations, and their application to the study of commutative number fields and their arithmetics.

In the first epoch (1907–19), Noether dealt primarily with differential and algebraic invariants, beginning with her dissertation under Paul Gordan. Her mathematical horizons broadened, and her work became more general and abstract, as she became acquainted with the work of David Hilbert, through close interactions with a successor to Gordan, Ernst Sigismund Fischer. After moving to Gottingen in 1915, she produced her seminal work for Physics, the two Noether's theorems.

In the second epoch (1920–26), Noether devoted herself to developing the theory of mathematical rings.

In the third epoch (1927–35), Noether focused on noncommutative algebra, linear transformations, and commutative number fields.

Mathematicians of previous centuries had worked on practical methods for solving specific types of equations, e.g., cubic, quartic, and quintic equations, as well as on the related problem of constructing regular polygons using compass and straightedge.

Commutative rings, ideals, and modules

Noether's paper, Ideal theory in Ringbereichen (Theory of Ideals in Ring Domains, 1921), is the foundation of general commutative ring theory, and gives one of the first general definitions of a commutative ring.

Elimination theory

In 1923–24, Noether applied her ideal theory to elimination theory showing that fundamental theorems about the factorization of polynomials could be carried over directly.

In a letter to The New York Times, Albert Einstein wrote

"In the judgment of the most competent living mathematicians, Noether was the most significant creative mathematical genius thus far produced since the higher education of women began."

Mathematician Norbert Wiener wrote

"Miss Noether is... the greatest woman mathematician who has ever lived; and the greatest woman scientist of any sort now living, and a scholar at least on the plane of Madame Curie."

Memorials

- The Association for Women in Mathematics holds a Noether Lecture to honour women in Mathematics every year; in its 2005 pamphlet for the event, the Association characterizes Noether as "one of the great mathematicians of her time, someone who worked and struggled for what she loved and believed in. Her life and work remain a tremendous inspiration".
- Consistent with her dedication to her students, the University of Siegen houses its Mathematics and Physics departments in buildings on the Emmy Noether Campus.
- A Street in her hometown, Erlangen, has been named after Emmy Noether and her father, Max Noether.
- The successor to the secondary school she attended in Erlangen has been renamed as the Emmy Noether School.
- A series of high school workshops and competitions are held in her honour in May of each year since 2001, originally hosted by a subsequent woman Mathematics Privatdozent of the University of Gottingen.
- Perimeter Institute for Theoretical Physics annually awards Emmy Noether Visiting Fellowships to outstanding female theoretical physicists

Fame after her Death

- The crater Nöther on the far side of the Moon is named after her.
- The 7001 Noether asteroid also is named for Emmy Noether.
- Google put a memorial doodle on Google's homepage in many countries on 23 March 2015 to celebrate Emmy Noether's 133rd birthday.

Emmy Noether Timeline

1882 : Noether was born on 23 March.
1900 : She taught English and French and attended Mathematics classes from the University of Erlangen.
1903 : She joined the University of Gottingen.
1907 : She received her Doctorate degree in Mathematics from University of Erlangen.
1915 : She joined the Mathematics department at the University of Gottingen.

1921 : She published 'Theory of Ideals in Ring Domains'.
1932 : She was awarded the Ackermann-Teubner Memorial Prize in Mathematics.
1933 : She became a guest professor at Bryn Mawr College in Pennsylvania, U.S.
1935 : Noether died on 14 April, from the complications that followed a uterine surgery.

Despite her intellectual achievements and the recognition of such mathematicians as David Hilbert and Hermann Weyl, Emmy Noether endured years of poor treatment by German universities, where for a time she could not even lecture under her own name. Weyl later wrote that, even when the Nazis prevented her from lecturing, "her courage, her frankness, her unconcern about her own fate, her conciliatory spirit, were, in the midst of all the hatred and meanness, despair and sorrow... a moral solace." Forced out of Germany by the Nazis in 1933, Emmy Noether came to Bryn Mawr College, where she soon collected many students and colleagues around her. She died there just two years later at the age of fifty-three.

Chapter 48

Gottfried Wilhelm Leibniz

God assuredly always chooses the best.

Born	:	**July 1, 1646, Leipzig**
Died	:	**November 14, 1716, Hannover, Germany**
Nationality	:	**German**

Gottfried Wilhelm Von Leibniz, a German mathematician and philosopher, was born July 1, 1646 in Leipzig, Germany. At age 15, he enrolled at the University of Leipzig, where his father used to teach, and earned both a bachelor's and master's degree in philosophy. He received his Doctor of Law from the University of Altdorf. He was a political advisor, important both as a metaphysician and as a logician and distinguished also for his independent invention of the differential and integral calculus.

Gottfried Leibniz discovered infinitesimal calculus, a distinction he shared with Sir Isaac Newton; however; each one made this discovery alone, not while working together. Infinitesimal calculus is the branch of Mathematics that is concerned with differentiation, integrations, and limits of functions. Leibniz also developed the law of continuity and transcendental law of homogeneity, cutting-edge theories that were not published or used in Mathematics until the 20th century.

Gottfried Leibniz was the first person to provide a description of a pinwheel calculator and invented a device of his own. The Leibniz wheel consisted of a cylinder with a set of teeth of increasing lengths. It was used in automatic mechanical calculators until the development of the electronic calculator in the mid-1970s.

Philosophical Principles

Leibniz variously invoked seven fundamental philosophical principles:

- Identity/contradiction: If a proposition is true, then its negation is false and vice versa.
- Identity of indiscernibles: Two distinct things cannot have all their properties in common.
- Sufficient reason. "There must be a sufficient reason for anything to exist, for any event to occur, for any truth to obtain."
- Pre-established harmony. "The appropriate nature of each substance brings it about that what happens to one corresponds to what happens to all the others, without, however, their acting upon one another directly
- Law of Continuity.
- Optimism. "God assuredly always chooses the best."
- Plenitude. Leibniz believed that the best of all possible worlds would actualize every genuine possibility.

Leibniz thought symbols were important for human understanding. He attached so much importance to the development of good notations that he attributed all his discoveries in Mathematics to this. His notation for the calculus is an example of his skill in this regard. Complex thoughts would be represented by combining characters for simpler thoughts.

Career

Leibnitz was only twenty one years old and he had already written several essays. The last essay which he wrote gathered much appreciation not only because he attempted reconstructing Corpus Juris, but also due to the fact that he bought forward the clear understanding of the importance of the historical method used in the law. Another relevant achievement of Leibniz was that he acquired the chance to meet Johann Christian von Boyneburg. Von Boyneburg employed Leibniz as his assistant and also introduced him to the elector of Mainz. As a step towards attaining employment, Leibniz wrote an essay for the Elector, following which he was asked to assist the redrafting for the Electorate, the legal code. In the year 1669, Leibniz was appointed in the Court of Appeal as Assessor. After von Boyneburg passed away in 1672, he was employed under his widow, who later dismissed him in 1674.

Contribution to Mathematics

- Leibniz can be considered the founder of that part of Mathematics.
- Bertrand Russell believe Leibniz developed calculus independently Although the mathematical notion of function was implicit in trigonometric and logarithmic tables, which existed in his day, Leibniz was the first, in 1692 and 1694, to employ it explicitly, to denote any of several geometric concepts derived from a curve, such as abscissa, ordinate, tangent, chord, and the perpendicular.
- Leibniz was the first to see that the coefficients of a system of linear equations could be arranged into an array, now called a matrix, which can be manipulated to find the solution of the system, if any. This method was later called Gaussian elimination.
- Leibniz's discoveries of Boolean algebra and of symbolic logic, also relevant to Mathematics, are discussed in the preceding section.
- Leibniz is credited, along with Sir Isaac Newton, with the discovery of calculus (differential and integral calculus). The product rule of differential calculus is still called "Leibniz's law". In addition, the theorem that tells how and when to differentiate under the integral sign is called the Leibniz integral rule.
- Leibniz exploited infinitesimals in developing the calculus, manipulating them in ways suggesting that they had paradoxical algebraic properties.

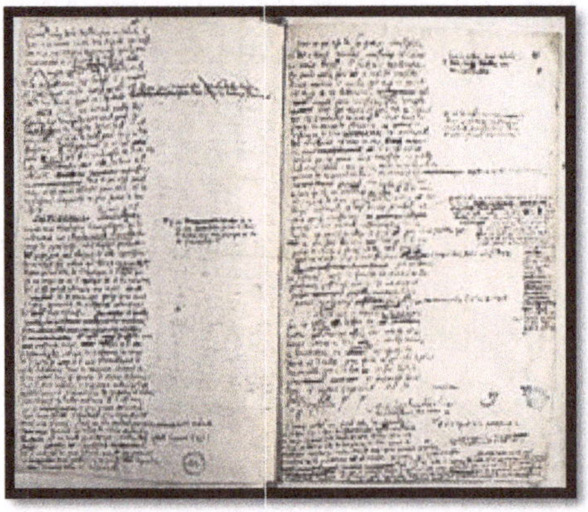

Leibniz's papers and notes from the National Library of Poland

Principal Works

- *De Arte Combinatoria* ('On the Art of Combination'), 1666
- *Hypothesis Physica Nova* ('New Physical Hypothesis'), 1671
- *Discours de métaphysique* ('Discourse on MetPhysics'), 1686
- unpublished manuscripts on the calculus of concepts, c. 1690
- *Nouveaux Essais sur L'entendement humaine* ('New Essays on Human Understanding'), 1705
- *Théodicée* ('Theodicy'), 1710
- *Monadologia* ('The Monadology'), 1714

Later Years

Leibniz then served the Brunswick family for the next forty years, under three successive princes. Thus, Leibniz moved over to a political surrounding which was formed by the dynastic aims of the German state. During this time, Leibniz devoted his time to his intellectual pursuits such as logic, Physics, philosophy, perfecting calculus and writing about other Mathematics related topics. In 1674, he started working on calculus and by 1677; he had made a coherent system with him, which he published only in 1684. Later the publication of papers in a journal between 1682-1692 enhanced his mathematical and scientific reputation. Leibniz was also appointed by the Elector Ernest Augustus to write the history of the House of Brunswick in an effort to enhance his dynastic ambitions. Thus, he had to go to Germany, Italy and Austria in order to find the archival materials related to this project from 1687-1690. Leibniz suffered accusations in 1708, by John Keil who wrote in the Royal Society journal of Leibniz having plagiarized Newton's works. The last thirty years of his Leibniz's life was busy with matters such as Mathematics, theology, history, jurisprudence, politics, science and philosophy.

Death

Gottfried W. Leibniz died on 14 November 1716 because of deteriorated health. His final days were filled with controversies and only his secretary attended the funeral. His grave was not marked for more than fifty years.

Leibniz's Life

- Born July 1, 1646, in Leipzig
- 1661, entered University of Leipzig (as a law student)
- 1663, baccalaureate thesis, *De Principio Individui* ('On the Principle of the Individual')
- 1667, entered the service of the Baron of Boineburg
- 1672 - 1676, lived in Paris (met Malebranche, Arnauld, Huygens)
- 1675, laid the foundation of the differential/integral calculus
- 1676, entered the service of the Duke of Hannover; worked on hydraulic presses, windmills, lamps, submarines, clocks, carriages, water pumps, the binary number system
- published *Nova Methodus Pro Maximus et Minimus* ('New Method for the Greatest and the Least'), an exposition of his differential calculus
- 1685, took on the duties of historian for the House of Brunswick
- 1691, named librarian at Wolfenbuettel
- 1700, named foreign member of the French Academy of Sciences in Paris
- 1711, met the Russian czar Peter the Great
- Died, November 14, 1716, in Hannover

Chapter 49

Karl Theodor Wilhelm Weierstrass

"Father of Modern Analysis"

Born : **October 31, 1815, Ennigerloh**
Died : **February 19, 1897, Province of Brandenburg**
Nationality : **German**

Biography

Weierstrass was the son of Wilhelm Weierstrass, a government official, and Theodora Vonderforst. His interest in Mathematics began while he was a *Gymnasium* student at Theodorianum in Paderborn. He was sent to the University of Bonn upon graduation to prepare for a government position. Because his studies were to be in the fields of law, economics, and finance, he was immediately in conflict with his hopes to study Mathematics. He resolved the conflict by paying little heed to his planned course of study, but continued private study in Mathematics. The outcome was to leave the university without a degree. After that he studied Mathematics at the University of Munster and his father was able to obtain a place for him in a teacher training school in Munster. Later he was certified as a teacher in that city. During this period of study, Weierstrass attended the lectures of Christoph Gudermann and became interested in elliptic functions.

Weierstrass may have had an illegitimate child named Franz with the widow of his friend Carl Wilhelm Borchardt.

After 1850 Weierstrass suffered from a long period of illness, but was able to publish papers that brought him fame and distinction. The University of Konigsberg conferred an honourary doctor's degree on him on 31 March 1854. In 1856 he took a chair at the Gewerbeinstitut, which later became the Technical University of Berlin. In 1864 he became professor at the Friedrich-Wilhelm's-Universität Berlin, which later became the Humboldt Universität zu Berlin. He was immobile for the last three years of his life, and died in Berlin from pneumonia.

Mathematical contributions

- **Soundness of calculus**

Weierstrass was interested in the soundness of calculus, and at the time, there were somewhat ambiguous definitions regarding the foundations of calculus, and hence important theorems could not be proven with sufficient rigour.

Delta-epsilon proofs are first found in the works of Cauchy in the 1820s. Cauchy did not clearly distinguish between continuity and uniform continuity on an interval. Notably, in his 1821 *Course d'analyse,* Cauchy argued that the (point wise) limit of (point wise) continuous functions was itself (point wise) continuous, a statement interpreted as being incorrect by many scholars. The correct statement is rather that the *uniform* limit of continuous functions is continuous (also, the uniform limit of uniformly continuous functions is uniformly continuous). This required the concept of uniform convergence, which was first observed by Weierstrass's advisor, Christophe Gudermann, in an 1838 paper, where Gudermann noted the phenomenon but did not define it or elaborate on it. Weierstrass saw the importance of the concept, and both formalized it and applied it widely throughout the foundations of calculus.

The formal definition of continuity of a function, as formulated by Weierstrass, is as follows:

$f(x)$ is continuous at $x = x_0$ if $\forall \in > \exists \delta > 0$ such that for every x in the domain of f,

$|x - x_0| < \delta \Rightarrow |f(x) - f(x_0)| < \varepsilon$.

Using this definition , he proved the Intermediate Value Theorem. He also proved the Bolzano-Weierstrass theorem and used it to study the properties of continuous functions on closed and bounded intervals.

- **Calculus of variations**

Weierstrass also made significant advancements in the field of calculus of variations. Using the apparatus of analysis that he helped to develop, Weierstrass was able to give a complete reformulation of the theory which paved the way for the modern study of the calculus of variations. Among the several significant axioms, Weierstrass established a necessary condition for the existence of strong extrema of variational problems. He also helped devise the Weierstrass–Erdmann condition, which gives sufficient conditions for an extremal to have a corner along a given extrema, and allows one to find a minimizing curve for a given integral.

Honours and awards

- Fellow of the Royal Society 1881
- Royal Society Copley Medal 1895
- The lunar crater Weierstrass and the asteroid 14100 Weierstrass are named after him.
- Also, there is the Weierstrass Institute for Applied Analysis and Stochastics in Berlin.

Quotation

"It is true that a mathematician who is not also something of a poet will never be a perfect mathematician.' – Karl Weierstrass

Chapter 50

Cantor

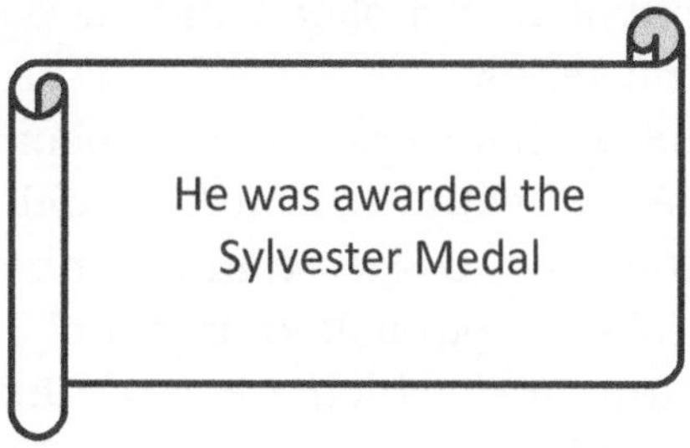

Born	:	**March 3, 1845, Saint Petersburg, Russian Empire**
Died	:	**January 6, 1918, Halle, Province of Saxony,German Empire**
Nationality	:	**German**

Youth and studies

Georg Cantor was born in the western merchant colony in Saint Petersburg, Russia, and brought up in the city until he was eleven. Georg, the oldest of six children, was regarded as an outstanding violinist. His grandfather Franz Bohm (1788–1846) (the violinist Joseph Bohm's brother) was a well-known musician and soloist in a Russian imperial orchestra. Cantor's father had been a member of the Saint Petersburg stock exchange; when he became ill, the family moved to Germany in 1856, first to Wiesbaden then to Frankfurt, seeking winters milder than those of Saint Petersburg.

In 1860, Cantor graduated with distinction from the Real schule in Darmstadt; his exceptional skills in Mathematics, trigonometry in particular, were noted. In 1862, Cantor entered the University of Zürich. After receiving a substantial inheritance upon his father's death in 1863, Cantor shifted his studies to the University of Berlin, attending lectures by Leopold Kronecker, Karl

Weierstrass and Ernst Kummer. He spent the summer of 1866 at the University of Gottingen, then and later a centre for mathematical research.

Career

Georg Cantor began his career by teaching in a Berlin girls' school for a brief period of time. He then accepted a position at the University of Halle, where he spent his entire career. He joined as a lecturer in 1869 and was promoted to assistant professor in 1872, and full professor in 1879.

He published a series of ten papers from 1869 to 1873 in which he dealt with the theory of numbers. A colleague, Heinrich Eduard Heine, recognized Cantor's capability and encouraged him to work on the theory of trigonometric series.

He started with the work performed on trigonometric series by the German mathematician Bernhard Riemann a few years ago, and extended it to show that the function of a complex variable can be represented in only one way by a trigonometric series.

He began his work on what became known as set theory in 1874. Cantor's correspondence with Richard Dedekind, mathematician at the Brunswick Technical Institute, triggered in his mind ideas on the theory of sets.

In 1874, he published an article 'On a Property of the Collection of All Real Algebraic Numbers' which marked the beginning of set theory as a branch of Mathematics. The article provided a rigorous proof that there was more than one kind of infinity. Through this work he proved that real numbers are not countable.

The article was seminal in more than one way. It also contained a new method of constructing transcendental numbers which were first constructed by Joseph Liouville in 1844.

Extending his research, he published a series of six articles in 'Mathematische Annalen' between 1879 and 1884 that formed an introduction to his set theory. He also introduced the Cantor set during this period. During this time he faced severe criticism of his works which is believed to have affected his mental health even though he continued his mathematical work despite the criticism.

In 1891, he published a paper containing his "diagonal argument" for the existence of an uncountable set. He applied the same idea to prove what is now known as Cantor's theorem. The theorem states that, for any set A, the set of all subsets of A has a strictly greater cardinality than A itself.

By this time his mental health had begun to deteriorate though he still determinedly performed mathematical research. His last significant papers on set theory were published in 1895 and 1897 in 'Mathematische Annalen' under Felix Klein's editorship.

Mathematical work

Cantor's work between 1874 and 1884 is the origin of set theory. Prior to this work, the concept of a set was a rather elementary one that had been used implicitly since the beginning of Mathematics, dating back to the ideas of Aristotle. No one had realized that set theory had any nontrivial content. Before Cantor, there were only finite sets (which are easy to understand) and "the infinite" (which was considered a topic for philosophical, rather than mathematical, discussion). By proving that there are (infinitely) many possible sizes for infinite sets, Cantor established that set theory was not trivial, and it needed to be studied. Set theory has come to play the role of a foundational theory in modern Mathematics, in the sense that it interprets propositions about mathematical objects (for example, numbers and functions) from all the traditional areas of Mathematics (such as algebra, analysis and topology) in a single theory, and provides a standard set of axioms to prove or disprove them. The basic concepts of set theory are now used throughout Mathematics.

In one of his earliest papers, Cantor proved that the set of real numbers is "more numerous" than the set of natural numbers; this showed, for the first time, that there exist infinite sets of different sizes. He was also the first to appreciate the importance of one-to-one correspondences (hereinafter denoted "1-to-1 correspondence") in set theory. He used this concept to define finite and infinite sets, subdividing the latter into denumerable (or countably infinite) sets and uncountable sets (nondenumerable infinite sets).

Cantor developed important concepts in topology and their relation to cardinality. For example, he showed that the Cantor set is nowhere dense, but has the same cardinality as the set of all real numbers, whereas the rational are everywhere dense, but countable.

Cantor introduced fundamental constructions in set theory, such as the power set of a set *A*, which is the set of all possible subsets of *A*. He later proved that the size of the power set of *A* is strictly larger than the size of *A*, even when *A* is an infinite set; this result soon became known as Cantor's theorem. Cantor developed an entire theory and arithmetic of infinite sets, called cardinals and ordinals, which extended the arithmetic of the natural numbers. His notation for the cardinal numbers was the Hebrew letter $\aleph$ (aleph) with a natural number subscript; for the ordinals he employed the Greek letter ω (omega). This notation is still in use today.

The *Continuum hypothesis*, introduced by Cantor, was presented by David Hilbert as the first of his twenty-three open problems in his address at the 1900 International Congress of Mathematicians in Paris. Cantor's work also attracted favourable notice beyond Hilbert's celebrated encomium.[15] The US philosopher Charles Sanders Peirce praised Cantor's set theory, and, following public lectures delivered by Cantor at the first International Congress of Mathematicians, held in Zurich in 1897, Hurwitz and Hadamard also both expressed their admiration. At that Congress, Cantor renewed his friendship and correspondence with Dedekind. From 1905, Cantor corresponded with his British admirer and translator Philip Jourdain on the history of set theory and on Cantor's religious ideas. This was later published, as were several of his expository works.

Number theory, trigonometric series and ordinals

Cantor's first ten papers were on number theory, his thesis topic. At the suggestion of Eduard Heine, the Professor at Halle, Cantor turned to analysis. Heine proposed that Cantor solve an open problem that had eluded Peter Gustav Lejeune Dirichlet, Rudolf Lipschitz, Bernhard Riemann, and Heine himself: the uniqueness of the representation of a function by trigonometric series. Cantor solved this difficult problem in 1869. It was while working on this problem that he discovered transfinite ordinals, which occurred as indices in the *n*th derived set S_n of a set S of zeros of a trigonometric series. Given a trigonometric series f(x) with S as its set of zeros, Cantor had discovered a procedure that produced another trigonometric series that had S_1 as its set of zeros, where S_1 is the set of limit points of S. If S_{k+1} is the set of limit points of S_k, then he could construct a trigonometric series whose zeros are S_{k+1}. Because the sets S_k were closed, they contained their Limit points, and the intersection of the infinite decreasing sequence of sets $S, S_1, S_2, S_3, \ldots$ formed a limit set, which we would now call S_ω, and then he noticed that S_ω would also have to have a set of limit points $S_{\omega+1}$, and so on. He had examples that went on forever, and so here was a naturally occurring infinite sequence of infinite numbers ω, $\omega + 1$, $\omega + 2$, ...

Between 1870 and 1872, Cantor published more papers on trigonometric series, and also a paper defining irrational numbers as convergent sequences of rational numbers. Dedekind, whom Cantor befriended in 1872, cited this paper later that year, in the paper where he first set out his celebrated definition of real numbers by Dedekind cuts. While extending the notion of number by means of his revolutionary concept of infinite cardinality, Cantor was paradoxically opposed to theories of infinitesimals of his contemporaries Otto Stolz and Paul du Bois-Reymond, describing them as both "an abomination" and "a cholera bacillus of Mathematics".[36] Cantor also published an erroneous "proof" of the inconsistency of infinitesimals.

Set theory

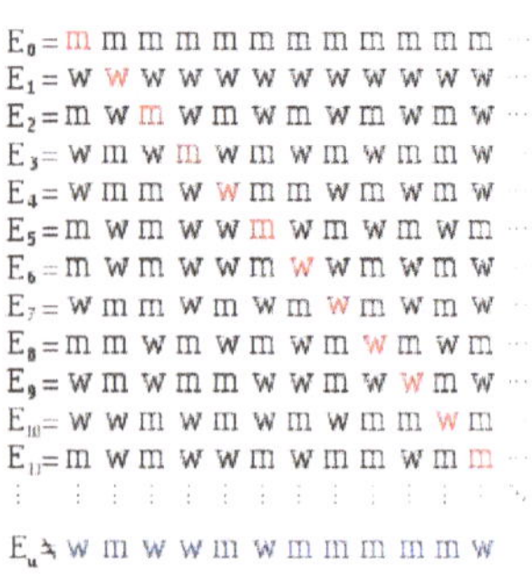

An illustration of Cantor's diagonal argument for the existence of uncountable sets. The sequence at the bottom cannot occur anywhere in the infinite list of sequences above.

The beginning of set theory as a branch of Mathematics is often marked by the publication of Cantor's 1874 article, *This article was the first to provide a rigorous proof that there was more than one kind of infinity.* Previously, all infinite collections had been implicitly assumed to be equinumerous (that is, of "the same size" or having the same number of elements). Cantor proved that the collection of real numbers and the collection of positive integers are not equinumerous. In other words, the real numbers are not countable. His proof differs from diagonal argument that he gave in 1891. Cantor's article also contains a new method of constructing transcendental numbers. Transcendental numbers were first constructed by Joseph Liouville in 1844.

Cantor established these results using two constructions. His first construction shows how to write the real algebraic numbers as a sequence a_1, a_2, a_3, ... In other words, the real algebraic numbers are countable. Cantor starts his second construction with any sequence of real numbers. Using this sequence, he constructs nested intervals whose intersection contains a real number not in the sequence. Since every sequence of real numbers can be used to construct a real not in the sequence, the real numbers cannot be written as a sequence – that is, the real numbers are not countable. By applying his construction to the sequence of real algebraic numbers, Cantor produces a transcendental number. Cantor points out that his constructions prove more – namely, they provide a new proof of Liouville's theorem: Every interval contains infinitely many transcendental numbers. Cantor's next article contains a construction that proves the set of transcendental numbers has the same "power" (see below) as the set of real numbers.

Between 1879 and 1884, Cantor published a series of six articles in *Mathematische Annalen* that together formed an introduction to his set theory. At the same time, there was growing opposition to Cantor's ideas, led by Kronecker, who admitted mathematical concepts only if they could be constructed in a finite number of steps from the natural numbers, which he took as intuitively given. For Kronecker, Cantor's hierarchy of infinities was inadmissible, since accepting the concept of actual infinity would open the door to paradoxes which would challenge the validity of Mathematics as a whole. Cantor also introduced the Cantor set during this period.

The fifth paper in this series, "Grundlagen einer allgemeinen Mannigfaltigkeitslehre" ("Foundations of a General Theory of Aggregates"), published in 1883, was the most important of the six and was also published as a separate monograph. It contained Cantor's reply to his critics and showed how the transfinite numbers were a systematic extension of the natural numbers. It begins by defining well-ordered sets. Ordinal numbers are then introduced as the order types of well-ordered sets. Cantor then defines the addition and multiplication of the cardinal and ordinal numbers. In 1885, Cantor extended his theory of order types so that the ordinal numbers simply became a special case of order types.

In 1891, he published a paper containing his elegant "diagonal argument" for the existence of an uncountable set. He applied the same idea to prove Cantor's theorem: the cardinality of the power set of a set *A* is strictly larger than the cardinality of *A*. This established the richness of the hierarchy of infinite sets, and of the cardinal and ordinal arithmetic that Cantor had defined. His argument is fundamental in the solution of the Halting problem and the proof of Gödel's first incompleteness theorem. Cantor wrote on the Gold batch conjecture in 1894.

In 1895 and 1897, Cantor published a two-part paper in *Mathematics Annalen* under Felix Klein's editorship; these were his last significant papers on set theory. The first paper begins by defining set, subset, etc., in ways that would be largely acceptable now. The cardinal and ordinal arithmetic are reviewed. Cantor wanted the second paper to include a proof of the continuum hypothesis, but had to settle for expositing his theory of well-ordered sets and ordinal numbers. Cantor attempts to prove that if *A* and *B* are sets with *A* equivalent to a subset of *B* and *B* equivalent to a subset of *A*, then *A* and *B* are equivalent. Ernst Schröder had stated this theorem a bit earlier, but his proof, as well as Cantor's, was flawed. Felix Bernstein supplied a correct proof in his 1898 PhD thesis; hence the name Cantor–Bernstein–Schroeder theorem.

One-to-one correspondence

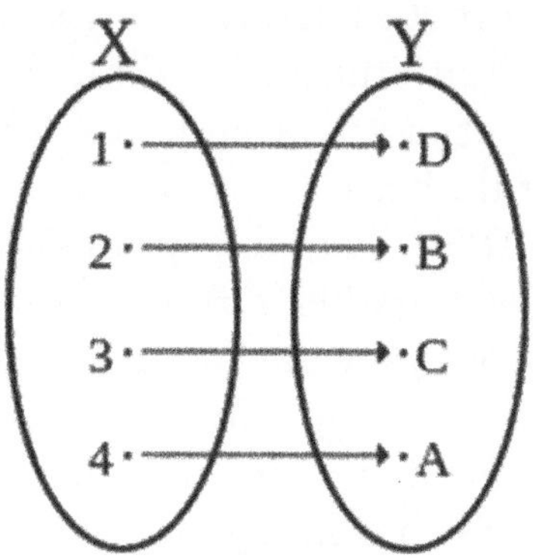

A bijective function

Cantor's 1874 Crelle paper was the first to invoke the notion of a 1-to-1 correspondence, though he did not use that phrase. He then began looking for a 1-to-1 correspondence between the points of the unit square and the points of a unit line segment. In an 1877 letter to Richard Dedekind, Cantor proved a far stronger result: for any positive integer *n*, there exists a 1-to-1 correspondence between the points on the unit line segment and all of the points in an *n*-dimensional space. *crois pas*!" The result that he found so astonishing has implications for geometry and the notion of dimension.

In 1878, Cantor submitted another paper to Crelle's Journal, in which he defined precisely the concept of a 1-to-1 correspondence, and introduced the notion of "power or "equivalence" of sets: two sets are equivalent (have the same power) if there exists a 1-to-1 correspondence between them. Cantor defined countable sets (or denumerable sets) as sets which can be put into a 1-to-1 correspondence with the natural numbers, and proved that the rational numbers are denumerable. He also proved that n-dimensional Euclidean space $\mathbf{R}^n$ has the same power as the real numbers $\mathbf{R}$, as does a countably infinite product of copies of $\mathbf{R}$. While he made free use of countability as a concept, he did not write the word "countable" until 1883. Cantor also discussed his thinking about dimension, stressing that his mapping between the unit interval and the unit square was not a continuous one.

This paper displeased Kronecker, and Cantor wanted to withdraw it; however, Dedekind persuaded him not to do so and Weierstrass supported its publication. Nevertheless, Cantor never again submitted anything to Crelle.

Awards and Achievements

The Royal Society awarded Cantor its Sylvester Medal, the highest honour it can confer for work in Mathematics, in 1904.

Personal Life and Legacy

- Georg Cantor married Vally Guttmann in 1874. It was a happy marriage and they had six children.
- He was prone to bouts of depression and suffered from mental illnesses during the later part of his life. His last years were spent in poverty and he spent the final year of his life in a sanatorium. He died on January 6, 1918.
- The Cantor Medal was established by the Deutsche Mathematiker-Vereinigung in honour of Georg Cantor.

Chapter 51

Niels Henrik Abel

> It appears to me that if one wishes to make progress in mathematics, one should study the masters and not the pupils.

Born : August 5, 1802, island of Finnøy, near Stavanger
Died : April 6, 1829, Froland
Nationality : Norwegian

Early life

Niels Henrik Abel was a Norwegian mathematician, born on 5th August 1802 in Nedstrand, Norway. Niels went to the Cathedral School at 13 in 1815. His teacher, Bernt Michael Holmboe recognized his talent in Mathematics and encouraged him to study the subject at the advance levels also tutoring Niels after school privately. Niels family underwent a severe crisis after 1818. Niels father died and as a consequence the family fell apart both emotionally and financially. Bernt Michael Holmboe helped Niels with a scholarship so he could stay in school and his friends raised enough money so Niels could apply to the Royal Frederick University.

Career

Christopher Hansteen's Help

After he graduated, professors from university supported Abel financially, and Professor Christopher Hansteen let him live in a room in the attic of his home. Abel would later view Ms. Hansteen as his second mother. While living here, Abel helped his younger brother, Peder Abel, through examen artium. He also helped his sister Elisabeth to find work in the town.

In early 1823, Niels Abel published his first article in "Magazine for Naturvidenskaberne", Norway's first scientific journal, which had been co-founded by Professor Hansteen. Abel published several articles, but the journal soon realized that this was not material for the common reader. In 1823, Abel also wrote a paper in French. It was "a general representation of the possibility to integrate all differential formulas" (*Norwegian: en alminnelig Fremstilling at Muligheten at integrere alle mulige Differential-Formler)*. He applied for funds at the university to publish it. However the work was lost, while being reviewed, never to be found thereafter.

Professor Rasmussen's Help

In mid-1823, Professor Rasmussen gave Abel a gift of 100 speciedaler so he could travel to Copenhagen and visit Ferdinand Degen and other mathematicians there. While in Copenhagen, Abel did some work on Fermat's Last Theorem. Abel's uncle, Peder Mandrup Tuxen, lived at the naval base in Christianshavn, Copenhagen, and at a ball there Niels Abel met Christine Kemp, his future fiancée. In 1824, Christine moved to Son, Norway to work as a governess and the couple got engaged over Christmas.

After returning from Copenhagen, Abel applied for a government scholarship in order to visit top mathematicians in Germany and France, but he was instead granted 200 speciedaler yearly for two years, to stay in Christiania and study German and French. In the next two years, he was promised a scholarship of 600 speciedaler yearly and he would then be permitted to travel abroad. While studying these languages, Abel published his first notable work in 1824, *Memoire sur les equations algebriques out on demote l'impossibilite de la resolution de l'equation generale du cinquieme degree* (Memoir on algebraic equations, in which the impossibility of solving the general equation of the fifth degree is proven). For, in 1823, Abel had at last proved the impossibility of solving the quintic equation in radicals (now referred to as the Abel–Ruffini theorem). However, this paper was in an abstruse and difficult form, in part because he had restricted himself to only six pages, in order to save money on printing. A more detailed proof was published in 1826 in the first volume of *Crelle's Journal.*

King Carl Johan of Norway/Sweden's Permission

In 1825, Abel wrote a personal letter to King Carl Johan of Norway/Sweden requesting permission to travel abroad. He was granted this permission, and in September 1825 he left Christiania together with four friends from university (Christian P.B Boeck, Balthazar M. Keilhau, Nicolay B. Moller and Otto Tank). These four friends of Abel were travelling to Berlin and to the Alps to study geology. Abel wanted to follow them to Copenhagen and from there make his way to Gottingen. The terms for his scholarship stipulated that he was to visit Gauss in Gottingen and then continue to Paris. However, when he got as far as Copenhagen he changed his plans. He wanted to follow his friends to Berlin instead, intending to visit Gottingen and Paris afterwards.

Meeting with Heinrich Christian Schumacher

On the way, he visited the astronomer Heinrich Christian Schumacher in Altona, now a district of Hamburg. He then spent four months in Berlin, where he became well acquainted with August Leopold Crelle, who was then about to publish his mathematical journal, Journal fur die reine und angewandte Mathematik. This project was warmly encouraged by Abel, who contributed much to the success of the venture. Abel contributed seven articles to it in its first year.

Abelian Functions

From Berlin Abel also followed his friends to the Alps. He went to Leipzig and Freiberg to visit Georg Amadeus Carl Friedrich Naumann and his brother the mathematician August Naumann. In Freiberg Abel did research in the theory of functions, particularly, elliptic, hyperelliptic, and a new class now known as abelian functions.

From Freiberg they went on to Dresden, Prague, Vienna, Trieste, Venice, Verona, Bolzano, Innsbruck, Luzern and Basel. From July 1826 Abel travelled on his own from Basel to Paris. Abel had sent most of his work to Berlin to be published in Crelle's Journal, but he had saved what he regarded as his most important work for the French Academy of Sciences, a theorem on addition of algebraic differentials.

With the help of Johan Gorlitz he found an apartment in Paris and continued his work on the theorem. He finished in October 1826, and submitted it to the academy. It was to be reviewed by Augustine-Louis Cauchy. Abel's work was scarcely known in Paris, and his modesty restrained him from proclaiming his research. The theorem was put aside and forgotten until his death.

Abel's limited finances finally compelled him to abandon his tour in January 1827. He returned to Berlin, and was offered a position as editor of Crelle's Journal, but opted out. By May 1827 he was back in Norway. His tour abroad was viewed as a failure. He had not visited Gauss in Gottingen and he had not published anything in Paris. His scholarship was therefore not renewed and he had to take up a private loan in Norges Bank of 200 spesidaler. He never repaid this loan. He also started tutoring. He continued to send most of his work to Crelle's Journal. But in mid-1828 he published, in rivalry with Carl Jacobi, an important work on elliptic functions in Astronomische Nachrichten in Altona.

Contributions to Mathematics

Quintic Equation

Abel showed that there is no general algebraic solution for the roots of a quintic equation, or any general polynomial equation of degree greater than four, in terms of explicit algebraic operations. To do this, he invented (independently of Galois) a branch of Mathematics known as group theory, which is invaluable not only in many areas of Mathematics, but for much of Physics as well. Abel sent a paper on the unsolvability of the quintic equation to Carl Friedrich Gauss, who proceeded to discard without a glance what he believed to be the worthless work of a crank.

Binomial Theorem

As a 16-year-old Abel gave a proof of the binomial theorem valid for all numbers, extending Euler's result which had held only for rationals. Abel wrote a fundamental work on the theory of elliptic integrals, containing the foundations of the theory of elliptic functions. While travelling to Paris he published a paper revealing the double periodicity of elliptic functions, which Adrien-Marie Legendre later described to Augustine-Louis Cauchy as "a monument more lasting than bronze" (borrowing a famous sentence by the Roman poetHoratius). The paper was however, misplaced by Cauchy.

While abroad Abel had sent most of his work to Berlin to be published in the Crelle's Journal, but he had saved what he regarded as his most important work for the French Academy of Sciences, a theorem on addition of algebraic differentials. The theorem was put aside and forgotten until his death. While in Freiberg Abel did research in the theory of functions, particularly, elliptic, hyper elliptic, and a new class now known as abelian functions.

Tragic Life

Niels Henrik Abel was a genius mathematician who lived a tragic life. When Abel sent his paper on elliptical functions to Gauss, Gauss threw it away without even reading it thinking it was a joke. When Abel travelled to France in 1825, Legendre and Cauchy (also 2 very famous mathematicians) described Abel as a “monument more lasting than bronze.” In spite of this, Cauchy also misplaced Abel’s manuscript. In Berlin, Abel became friends with mathematician August Crelle. However, Abel was still not offered a post as professor, nor were his works widely acknowledged.

Unable to find a job and living in poverty, Abel died of tuberculosis in 1829 at only 26 years old. Tragically, a letter arrived to Abel from his friend August Crelle informing him that he had been offered a position as professor at the University of Berlin two days after Abel’s death, but it was already too late. Abel lived a short life filled with many amazing accomplishments. He was underappreciated due partially to the fact that he was Norwegian; he would have been more famous if he had come from France or Germany. Today an Abel’s Prize exists in Niels Abel’s honour. The term ‘abelian’ has been coined in several concepts. Hermite quotes, “Abel has left mathematicians enough to keep them busy for 500 years.

Legacy

- Under Abel's guidance new fields were entered upon and the study of functions so advanced as to provide mathematicians with numerous ramifications along which progress could be made.
- His works, the greater part of which originally appeared in Crelle's Journal, were edited by Bernt Michael Holmboe and published in 1839 by the Norwegian government, and a more complete edition by Ludwig Sylow and Sophus Lie was published in 1881.
- The adjective "abelian", derived from his name, has become so common place in mathematical writing that it is conventionally spelled with a lower-case initial "a" (e.g., abelian group, abelian category, and abelian variety).
- On 6 April 1929, four Norwegian stamps were issued for the centenary of Abel's death.
- His portrait appears on the 500-kroner banknote (version V) issued during 1978–1985.
- On 5 June 2002, four Norwegian stamps were issued in honour of Abel two months before the bicentenary of his birth.
- There is also a 20-kroner coin issued by Norway in his honour.
- A statue of Abel stands in Oslo, and crater Abel on the Moon was named after him.
- In 2002, the Abel Prize was established in his memory.

Vinsha R

1st Year B.Ed Student

Chapter 52

Joseph-Louis Lagrange

His name is one of the 72 names inscribed on the Eiffel Tower

Born : January 25, 1736, Turin, Piedmont-Sardinia
Died : April 10, 1813, Paris, France
Nationality : Italian

Joseph Louis Lagrange, originally considered a French Mathematician, was actually of an Italian descent. He was largely self-taught and pushed himself to reach the top of the ladder during the early 18th century. One of the most proficient mathematicians at the time, Lagrange was responsible for developing many theories related to mechanics and also studied the number system. He studied a wide variety of topics and was a published astronomer who studied the solar system in great depths.

His consistent works in the field of radicals, permutations, and fluid mechanics made him a pioneer of his time. He was appreciated by Napoleon for his efforts and was awarded the Legion of Honour. Lagrange was also given the honour of becoming a member of the Academy of Sciences. With his wide range of admirers and an unbeatable passion for work, Joseph Louis Lagrange went on to become one of the first and definitely, one of the most influential faces in the history of Mathematics and science.

Childhood and Early Life

Joseph Louis Lagrange, named Giuseppe Lodovico Lagrangia, was born to Italian parents of French descent on 25th January 1736. Lagrange's father was the head of the Sardinian military chest and was at an influential position. Born into a wealthy family, Lagrange never cultivated an interest or taste for Mathematics. Although the family was in a wealthy spot, Lagrange's father's gambling habit led to turbulent times for the gamily. Lagrange was raised as a Catholic but he soon turned Agnostic when he had joined college. He discovered his passion for Mathematics at the age of 17 when he fortuitously came across a paper on the subject written by Edmund Halley.

Career

After a brief stint as a Mathematics professor in an artillery school, Lagrange decided to concentrate work on a specific division of Mathematics that interested him the most; calculus. He developed theories in 1754 and also engaged himself in the subject of classical mechanics. It was during this period that he discovered the functions of tautochrone and also generalized the concepts of Euler's analysis, after developing the Euler-Lagrange equations. Along with the help of his pupils in 1758, he established a guild called the Turin Academy of Sciences where he wrote elaborate papers and dissertations called the 'Miscellanea Taurinensia' and even discussed calculus at great lengths. In this paper, Lagrange pointed out a few errors made by Newton and also discussed theories on the recurring series, dynamics, integral calculus and probabilities. In 1764, he deliberated over the 'three body problem' and studied celestial bodies such as planet Jupiter in 1766. His passion to understand the dynamics of the solar system got him to discover various solutions that came to be known as the 'Lagrangian points'. He defied traditional science and contributed heavily to even 'Newtonian Mechanics' which were later renamed as 'Lagrangian Mechanics'.

Following his papers and discussions, King Frederick of Prussia persuaded Lagrange to reside at his court 'as the greatest mathematician in Europe'. It was here he produced his epic work, the 'Mecanique Analytique'. Lagrange became extremely popular with the Prussian King, and he enjoyed the space given to him, in terms of a regular life, where he could sit and effortlessly write his papers without any hindrance or discomfort. He was pleased with his tenure at the court, but it was soon followed by a tragedy, with the death of the King in 1786.

After the death of Frederick and at the behest of King Louis XVI, Lagrange moved to Paris where he was soon accorded a prestigious place at the "French Academy of Sciences". He lived a solitary life in Paris, and brooded over the French Revolution that took place. Boredom overtook his life to such an extent that he did not even dare open his fabled 'Mecanique Analytique' treatise, which lay on his work desk untouched.

Life started to get back on track in 1794 when Lagrange was appointed professor at the 'Ecole Polytechnique' where he was admired and venerated by his pupils. Towards the end of 1795, Lagrange and a few others went on to establish the 'Bureau des Longitudes' that was founded for the purpose astronomical observation and standard time keeping. He was given the honour of the 'Mathematical chair' at the new institution called 'Ecole Normale'. After much deliberation and research, Lagrange's theory on the decimal subdivision was finally accepted by the French commission in 1799.

Later Years

Prior to his death in 1810, Lagrange began to correct and revise the "Mecanique Analytique" but couldn't complete the task. Towards the end of the 1790's he made noteworthy discoveries in the field of continued fractions, interpolation and another treatise called the 'Lecons sur le calcul des fonctions' which studied the various aspects of mechanics, calculus and the applications to Geometry. Just before his death, Lagrange was made the Legion officer by Napoleon, and was also awarded the "Grand Croix" of the 'Ordre Imperial de la Reunion' in the year 1813.

Personal Life

Joseph Louis Lagrange married twice in his lifetime, though both his marriages were short lived. He married his first wife Vittoria Conti, who was also his first cousin, in 1767. The marriage was rather bitter and, as a result, didn't bear any children. After the death of his first wife, Lagrange married Renee-Francoise-Adelaide Le Monnier, daughter of his colleague, in 1792 and led a happy married life.

Death and Legacy

Joseph Louis Lagrange died on April 10 1813, in Paris just a week after he was awarded the 'Grande Croix'. He left behind a profound legacy and his contributions to the subjects of Mathematics and astronomy paved way for popular students, such as Giovanni Plana, Simeon Poisson and Joseph Fourier, who carried his epic works into the next generation. He left a mark so deep that his name is one of the 72 names inscribed on the Eiffel Tower.

Books by Joseph Louis Lagrange

- Lectures On Elementary Mathematics
- Analytische Mechanik ...: Aus Dem Französischen Mit Einigen Anmerkungen Und Erlauternden Zusätzen (German Edition)
- J. L. Lagrange's Mathematische Werke, Volume 1... (German Edition)

Prizes and Distinctions

- Euler proposed Lagrange for election to the Berlin Academy and he was elected on 2 September 1756.
- He was elected a Fellow of the Royal Society of Edinburgh in 1790, a Fellow of the Royal Society and a foreign member of the Royal Swedish Academy of Sciences in 1806.

- In 1808, Napoleon made Lagrange a Grand Officer of the Legion of Honour and a Count of the Empire.
- He was awarded the Grand Croix of the Ordre Impérial de la Réunion in 1813, a week before his death in Paris.
- Lagrange was awarded the 1764 prize of the French Academy of Sciences for his memoir on the libration of the Moon.
- In 1766 the Academy proposed a problem of the motion of the satellites of Jupiter, and the prize again was awarded to Lagrange.
- He also shared or won the prizes of 1772, 1774, and 1778.
- Lagrange is one of the 72 prominent French scientists who were commemorated on plaques at the first stage of the Eiffel Tower when it first opened.
- *Rue Lagrange* in the 5th Arrondissement in Paris is named after him.
- In Turin, the street where the house of his birth still stands is named *via Lagrange.*
- The lunar crater Lagrange also bears his name.

Chapter 53

George Boole

"No matter how correct a mathematical theorem may appear to be, one ought never to be satisfied that there was not something imperfect about it until it also gives the impression of being beautiful"

Born : November 2, 1815, Lincoln, Lincolnshire, England
Died : December 8, 1864, Ballintemple, Cork, Ireland
Nationality : British

Childhood and Early Life

George Boole was born in Lincoln, Lincolnshire, England, on 2nd November 1815. His parents were John Boole, a shoemaker, and Mary Ann Joyce. He had three younger siblings. He received his early education in a primary school and was also tutored in Mathematics by his father. An intelligent boy, he did not get any higher education but was largely self-taught in Mathematics and modern languages.

Career

Due to dire family circumstances he was forced to become the family's breadwinner at the age of 16 and took up a junior teaching position in Doncasterat Heigham's School. He also taught at some other schools.

During this period he furthered his mathematical knowledge by participating in the local Mechanics Institute, the Lincoln Mechanics' Institution. There he was given the calculus text of Sylvestre François Lacroix by the Rev. George Stevens Dickson of St Swithin's, Lincoln. He studied calculus by himself for years and eventually mastered it.

An extremely hard working young man, he established his own school in Lincoln when he was 19. Four years later, he took over Hall's Academy in Waddington, outside Lincoln.

In 1839, he submitted several papers like 'Researches on the Theory of Analytical Transformations' to the 'Cambridge Mathematical Journal'. These papers discussed differential equations and algebraic problem of linear transformation by emphasizing the idea of invariance linear transformation by emphasizing the idea of invariance.

In 1840, he returned to Lincoln to operate a boarding school. He discovered Invariant Theory, a new branch of Mathematics the next year. This particular mathematical theory was a source of inspiration for Einstein.

In 1844, he published an important paper in the 'Philosophical Transactions of the Royal Society' in which he discussed how methods of algebra and calculus might be combined.

Even though he never received a formal training in Mathematics, he developed new ideas on logical method and published a pamphlet, 'Mathematical Analysis of Logic,' in 1847 in which he argued that logic should be allied with Mathematics, not philosophy.

His publications were regarded very highly and earned him an appointment as professor of Mathematics at Queen's College, County Cork, in 1849 even though he had no university degree.

In 1854, he worked on the algebra of logic which is better known as Boolean algebra. He introduced general symbolic method of logical inference through 'Laws of Thought', published in the same year.

Boolean algebra acts as the basis of analyzing the validity of logical propositions as it captures the two-valued character of statements that may appear either true or false. This method of two-valued character and logical elements of Boolean logic is used in telephone switching and in electronic computers for their design and operation.

In 1857, he published 'On the Comparison of Transcendents , with Certain Applications to the Theory of Definite Integrals'. In it he discussed about the sum of residues of a rational function. As part of his discussion, he proved Boole's identity.

In 1859, he published 'Treatise on Differential Equations'. Through this work he informed about the general symbolic method. In the next year he published its sequel 'Treatise on the Calculus of Finite Differences'.

Death

In late November 1864, Boole walked, in heavy rain, from his home at Lichfield Cottage in Ballintemple to the university, a distance of three miles, and lectured wearing his wet clothes. He soon became ill, developing a severe cold and high fever. As his wife believed that remedies should resemble their cause, she put her husband to bed and poured buckets of water over him – the wet having brought on his illness. Boole's condition worsened and on 8 December 1864, he died of fever-induced pleural effusion.

He was buried in the Church of Ireland cemetery of St Michael's, Church Road, Blackrock (a suburb of Cork). There is a commemorative plaque inside the adjoining church.

Honours and awards

George Boole was awarded the Keith Medal by the Royal Society of Edinburgh in 1855.

In 1857, he was elected a Fellow of the Royal Society (FRS).

He is also the recipient of honourary degrees of LL.D. from the University of Dublin and the University of Oxford.

Legacy

Boolean algebra is named after him, as is the crater Boole on the Moon.

The keyword Bool represents a Boolean datatype in many programming languages, though Pascal and Java, among others, both use the full name Boolean.

The library, underground lecture theatre complex and the Boole Centre for Research in Informatics at University College Cork are named in his honour.

A road called Boole Heights in Bracknell, Berkshire is named after him.

Boole's work was extended and refined by a number of writers, beginning with William Stanley Jevons. Augustus De Morgan had worked on the logic of relations, and Charles Sanders Peirce integrated his work with Boole's during the 1870s. Other significant figures were Platon Sergeevich Poretskii, and William Ernest Johnson. The conception of a Boolean algebra structure on equivalent statements of a propositional calculus is credited to Hugh MacColl (1877), in work surveyed 15 years later by Johnson. Surveys of these developments were published by Ernst Schröder, Louis Couturat, and Clarence Irving Lewis.

In modern notation, the free Boolean algebra on basic propositions p and qarranged in a Hasse diagram. The Boolean combinations make up 16 different propositions, and the lines show which are logically related.

In 1921 the economist John Maynard Keynes published a book on probability theory, A Treatise of Probability. Keynes believed that Boole had made a fundamental error in his definition of independence which vitiated much of his analysis.[38] In his book The Last Challenge Problem, David Miller provides a general method in accord with Boole's system and attempts to solve the problems recognised earlier by Keynes and others. Theodore Hailperin showed much earlier that Boole had used the correct mathematical definition of independence in his worked out problems.

Boole's work and that of later logicians initially appeared to have no engineering uses. Claude Shannon attended a philosophy class at theUniversity of Michigan which introduced him to Boole's studies. Shannon recognised that Boole's work could form the basis of mechanisms and processes in the real world and that it was therefore highly relevant. In 1937 Shannon went on to write a master's thesis, at the Massachusetts Institute of Technology, in which he showed how Boolean algebra could optimise the design of systems of electromechanical relays then used in telephone routing switches. He also proved that circuits with relays could solve Boolean algebra problems. Employing the properties of electrical switches to process logic is the basic concept that underlies all modern electronic digital computers.

Victor Shestakov at Moscow State University (1907–1987) proposed a theory of electric switches based on Boolean logic even earlier than Claude Shannon in 1935 on the testimony of Soviet logicians and mathematicians Sofya Yanovskaya, Gaaze-Rapoport,Roland Dobrushin, Lupanov, Medvedev and Uspensky, though they presented their academic theses in the same year, 1938. But the first publication of Shestakov's result took place only in 1941 (in Russian). Hence, Boolean algebra became the foundation of practical digital circuitdesign; and Boole, via Shannon and Shestakov, provided the theoretical grounding for the Information Age.

"UCC's George Boole 200 project, featured events, student outreach activities and academic conferences on Boole's legacy in the digital age, including a new edition of Desmond MacHale's 1985 biography The Life and Work of George Boole: A Prelude to the Digital Age (Cork University Press, 2014).

The search engine Google marked the 200th anniversary of his birth on 2 November 2015 with an algebraic reimaging of its Google Doodle.

Jemima Monisha G

1st Year B.Ed Student

Chapter 54

Rene Descartes

He is called the "Father of Analytical Geometry".

Born : March 31, 1596, in La Haye en Touraine, Touraine
Died : February 11, 1650, Stockholm, Sweden
Nationality : French

- If you would be a real seeker after truth, it is necessary that at least once in your life you doubt, as far as possible, all things.
- Except our own thoughts, there is nothing absolutely in our power.
- It is only prudent never to place complete confidence in that by which we have even once been deceived.

Early life

Rene Descartes was born on March 31, 1596, in La Haye, France. His father, Joachim, served in the Parliament of Brittany, France. Jeanne Brochard Descartes, his mother, died in 1597. His father remarried and Rene and his older brother and sister were raised by their maternal grandmother and by a nurse for whom he retained a deep affection. In 1606 Descartes entered La Fleche, a religious college established for the education of the sons of noblemen. As a child he was often ill and was allowed to spend a portion of each day studying in bed. He used this time for meditation and thought. According to Descartes's description of his eight-year course of studies at La Fleche, he often felt embarrassed at the extent of his own ignorance.

Contribution to Mathematics

Rene Descartes has been dubbed the "Father of Modern Philosophy", but he was also one of the key figures in the Scientific Revolution of the 17th Century, and is sometimes considered the first of the modern school of Mathematics.

As a young man, he found employment for a time as a soldier. But, after a series of dreams or visions, and after meeting the Dutch philosopher and scientist Isaac Beeckman, who sparked his interest in Mathematics and the New Physics, he concluded that his real path in life was the pursuit of true wisdom and science.

Back in France, the young Descartes soon came to the conclusion that the key to philosophy, with all its uncertainties and ambiguity, was to build it on the indisputable facts of Mathematics.

In mathematics, his contribution lies chiefly in geometry that's why today he is known as father of analytical geometry. His main achievement was to bridge the gulf between algebra and geometry. Thus he is widely acclaimed as first mathematician who laid the foundation of modern geometry that resulted in development of analysis and calculus. With regard to algebra, he explained in detail that how algebric equations can be expressed and explained through use of geometrical shapes.

He also sought to find out the systematic meaning of knowledge by application of mathematical techniques. In this way, he broke from conventional scholastic-Aristotelian philosophy that used to explain haphazardly the interrelationship between ideas. Thus, instead of following the tradition, he replaced this casual explanation of nature of things with more scientific method. He intended to reach at mechanistic rationalization of beliefs by providing a more concrete basis. This mechanistic principle is not only applied to human or animal bodies but also plants. Thus, he embarked upon a quest to engage in a scientific enquiry as contrary to his predecessors. He also changed the scholastic explanation of substantial forms in physics by his mechanistic philosophy.

Descartes also tended to apply geometric method to physics and also explained it by deductive method that results can be inferred by perceptions of geometric properties of body. As Descartes explained this "from what has already been said we have established that all the bodies in the universe are composed of one and the same matter, which is divisible into indefinitely many parts, and is in fact divided into a large number of parts which move in different directions and have a sort of circular motion; moreover, the same quantity of motion is always preserved in the universe"

His major contribution lies in bringing forth coordinate system that also bears his name. This Cartesian coordinate system tended to explain the algebraic equations through geometrical shapes. He "invented the convention of representing unknowns in equations by x, y and z". It was his work of calculus that was later used by Newton thus evolving a new branch of mathematics.

Besides that, he also invented rule of signs to establish the positive and negative roots of polynomial.

Other contribution

In his treatise "Discourses on Methods", he called into question scholastic beliefs and put forward a deductive method that implies inferring through generalization rather than sensation. As his famous dictum is Cogito, ergo sum or "I think, therefore I am."

He attempted to construct a system of knowledge based upon rationalization and logical deduction by negating the principle of perception. He believed that reason is the only way to attain higher knowledge. In optics, he also discovered laws of reflection and refraction. Rene Descartes also advanced his views on motion of objects in his treatise "Principles of Philosophy". His ideas influenced later philosophers like Hobbes, Pascal, Locke and Kant.

Death

He died on 11 February 1650 in Stockholm, Sweden due to pneumonia. He was then serving as a tutor for Queen of Sweden. He was laid to rest at the Abbey of Saint-Germain-des-Prés, Paris. A memorial of René Déscartes was built in the Swedish church in the eighteenth century.

Legacy

Descartes left rich legacy in mathematics by ideas on Cartesian geometry and creation of XYZ as representation for unknown equation. His works became foundation for development of calculus theory by Leibinz and Newton. Besides, he also made contribution in the field of optics.

Major Works

- Compendium Musicae (1618)
- Regulae ad directionem ingenii (Rules for the Direction of the Mind (1626–1628)
- Le Monde (The World) and L'Homme (Man)- 1630–1633.
- Discours de la méthode -Discourse on the Method (1637).
- La Géométrie (1637). Descartes's major work in mathematics.
- Meditationes de prima philosophia- Meditations on First Philosophy (1641)
- Principia philosophiae (1644)
- Les passions de l'âme- Passions of the Soul (1649.)
- Musicae Compendium- Instruction in Music (1656).

Chapter 55

Pascal

"The heart has its reasons which reason knows nothing of"

Born : June 19, 1623, Clermont-Ferrand, Auvergne
Died : August 19, 1662, Paris, France
Nationality : French

Early Life

Born at Clermont-Ferrand on June 19, 1623, Pascal's father was Etienne Pascal, a counselor of the king who later became the president of the Court of Aids at Clermont. His mother died in 1626. The Pascal family settled in Paris in 1631.

At a tender age of 12, Pascal began participating in the meetings of a mathematical academy. He learned different languages from his father, Latin and Greek in particular, but Pascal Sr. didn't teach him Mathematics. This increased the curiosity of young Pascal, who went on to experiment with geometrical figures, even formulating his own names for standard geometrical terms.

Contribution to Mathematics

The Frenchman Blaise Pascal was a prominent 17th Century scientist, philosopher and mathematician. Like so many great mathematicians, he was a child prodigy and pursued many different avenues of intellectual endeavour throughout his life. Much of his early work was in the area of natural and applied sciences, and he has a physical law named after him (that "pressure exerted anywhere in a confined liquid is transmitted equally and undiminished in all directions throughout the liquid"), as well as the international unit for the meaurement of pressure. In philosophy, Pascals' Wager is his pragmatic approach to believing in God on the grounds that is it is a better "bet" than not to.

But Pascal was also a mathematician of the first order. At the age of sixteen, he wrote a significant treatise on the subject of projective geometry, known as Pascal's Theorem, which states that, if a hexagon is inscribed in a circle, then the three intersection points of opposite sides lie on a single line, called the Pascal line. As a young man, he built a functional calculating machine, able to perform additions and subtractions, to help his father with his tax calculations. He is best known, however, for Pascal's Triangle, a convenient tabular presentation of binomial co-efficients, where each number is the sum of the two numbers directly above it. A binomial is a simple type of algebraic expression which has just two terms operated on only by addition, subtraction, multiplication and positive whole-number exponents, such as $(x + y)^2$. The co-efficients produced when a binomial is expanded form a symmetrical triangle (see image at right).

Pascal was far from the first to study this triangle. The Persian mathematician Al-Karaji had produced something very similar as early as the 10th Century, and the Triangle is called Yang Hui's Triangle in China after the 13th Century Chinese mathematician, and Tartaglia's Triangle in Italy after the eponymous 16th Century Italian. But Pascal did contribute an elegant proof by defining the numbers by recursion, and he also discovered many useful and interesting patterns among the rows, columns and diagonals of the array of numbers. For instance, looking at the diagonals alone, after the outside "skin" of 1's, the next diagonal (1, 2, 3, 4, 5,...) is the natural numbers in order. The next diagonal within that (1, 3, 6, 10, 15,...) is the triangular numbers in order. The next (1, 4, 10, 20, 35,...) is the pyramidal triangular numbers, etc, etc. It is also possible to find prime numbers, Fibonacci numbers, Catalan numbers, and many other series, and even to find fractal patterns within it.

When a binomial expression is raised to a power, the Binomial Theorem says that the co-efficients of the terms in the expansion of the binomial follow a distinct pattern.

For example:

$(x+y)^0 = 1$

$(x+y)^1 = x + y$

$(x+y)^2 = x^2 + 2xy + y^2$

$(x+y)^3 = x^3 + 3x^2y + 3xy^2 + y^3$

$(x+y)^4 = x^4 + 4x^3y + 6x^2y^2 + 4xy^3 + y^4$

$(x+y)^5 = x^5 + 5x^4y + 10x^3y^2 + 10x^2y^3 + 5xy^4 + y^5$

The co-efficients can be shown as a triangle where each number is the sum of the two above it

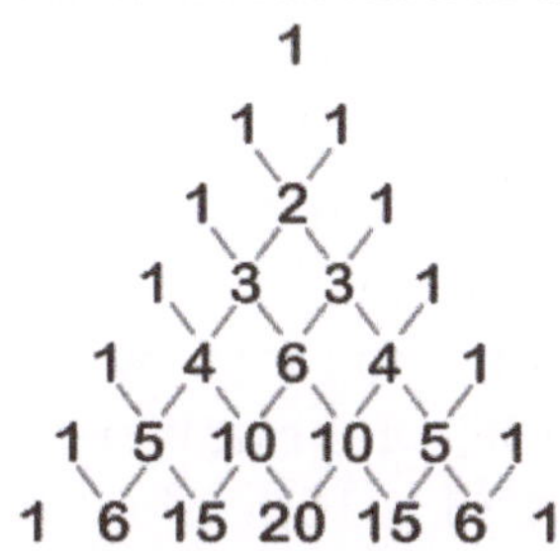

This is often known as Pascal's Triangle, although Al-Kharaji, Yang Hui and Tartaglia had all produced similar results centuries earlier.

Pascal also made the conceptual leap to use the Triangle to help solve problems in probability theory. In fact, it was through his collaboration and correspondence with his French contemporary Pierre de Fermat and the Dutchman Christiaan Huygens on the subject that the mathematical theory of probability was born. Before Pascal, there was no actual theory of probability - notwithstanding Gerolamo Cardano's early exposition in the 16th Century - merely an understanding (of sorts) of how to compute "chances" in dice and card games by counting equally probable outcomes. Some apparently quite elementary problems in probability had eluded some of the best mathematicians, or given rise to incorrect solutions.

It fell to Pascal (with Fermat's help) to bring together the separate threads of prior knowledge (including Cardano's early work) and to introduce entirely new mathematical techniques for the solution of problems that had hitherto resisted solution. Two such intransigent problems which Pascal and Fermat applied themselves to were the Gambler's Ruin (determining the chances of winning for each of two men playing a particular dice game with very specific rules) and the Problem of Points (determining how a game's winnings should be divided between two equally skilled players if the game was ended prematurely). His work on the Problem of Points in particular, although unpublished at the time, was highly influential in the unfolding new field.

The Problem of Points at its simplest can be illustrated by a simple game of "winner take all" involving the tossing of a coin. The first of the two players (say, Fermat and Pascal) to achieve ten points or wins is to receive a pot of 100 francs. But, if the game is interrupted at the point where Fermat, say, is winning 8 points to 7, how is the 100 franc pot to divided? Fermat claimed that, as he needed only two more points to win the game, and Pascal needed three, the game would have been over after four more tosses of the coin (because, if Pascal did not get the necessary 3 points for your victory over the four tosses, then Fermat must have gained the necessary 2 points for his victory, and vice versa. Fermat then exhaustively listed the possible outcomes of the four tosses, and concluded that he would win in 11 out of the 16 possible outcomes, so he suggested that the 100 francs be split $^{11}/_{16}$ (0.6875) to him and $^{5}/_{16}$ (0.3125) to Pascal.

Pascal then looked for a way of generalizing the problem that would avoid the tedious listing of possibilities, and realized that he could use rows from his triangle of coefficients to generate the numbers, no matter how many tosses of the coin remained. As Fermat needed 2 more points to win the game and Pascal needed 3, he went to the fifth (2 + 3) row of the triangle, i.e. 1, 4, 6, 4, 1. The first 3 terms added together (1 + 4 + 6 = 11) represented the outcomes where Fermat would win, and the last two terms (4 + 1 = 5) the outcomes where Pascal would win, out of the total number of outcomes represented by the sum of the whole row (1 + 4 + 6 +4 +1 = 16).

If four coins are to be tossed to finish a game, the possible outcomes are:

HHHH	HHHT	HHTH	HHTT
HTHH	HTHT	HTTH	HTTT
THHH	THHT	THTH	THTT
TTHH	TTHT	TTTH	TTTT

The 11 yellow outcomes are ones where Fermat, who only needs 2 more points, wins the game, and the 5 blue outcomes are ones where Pascal wins.

Therefore, Fermat should receive $^{11}/_{16}$ of the pot and Pascal should receive $^{5}/_{16}$.

Fermat and Pascal's solution to the Problem of Points

Pascal and Fermat had grasped through their correspondence a very important concept that, though perhaps intuitive to us today, was all but revolutionary in 1654. This was the idea of equally probable outcomes, that the probability of something occurring could be computed by enumerating the number of equally likely ways it could occur, and dividing this by the total number of possible outcomes of the given situation. This allowed the use of fractions and ratios in the calculation of the likelihood of events, and the operation of multiplication and addition on these fractional probabilities. For example, the probability of throwing a 6 on a die twice is $\frac{1}{6} \times \frac{1}{6} = \frac{1}{36}$ ("and" works like multiplication); the probability of throwing either a 3 or a 6 is $\frac{1}{6} + \frac{1}{6} = \frac{1}{3}$ ("or" works like addition).

Later Life

Later in life, Pascal and his sister Jacqueline strongly identified with the extreme Catholic religious movement of Jansenism. Following the death of his father and a "mystical experience" in late 1654, he had his "second conversion" and abandoned his scientific work completely, devoting himself to philosophy and theology. His two most famous works, the "Lettres provinciales" and the "Pensées", date from this period, the latter left incomplete at his death in 1662. They remain Pascal's best known legacy, and he is usually remembered today as one of the most important authors of the French Classical Period and one of the greatest masters of French prose, much more than for his contributions to Mathematics.

Conclusion

Mathematics is undoubtedly a fascinating subject of study. It enthused a host of creative people to touch new heights by applying Mathematics to varied fields of human knowledge and interest, and offer bewildering facts about new fields of learning and research, useful for all across the world.

Interesting Facts

Blaise Pascal only lived for 39 years, but they were 39 years full of discoveries, learning, and creativity. Born on June 19, 1623 in France, Blaise Pascal lost his mother only three years after in 1626. He died himself on August 19, 1662 in Paris.

Legacy

In honour of his scientific contributions, the name **Pascal** has been given to the SI unit of pressure, to a programming language, andPascal's law (an important principle of hydrostatics), and as mentioned above, Pascal's triangle and Pascal's wager still bear his name.

Pascal's development of probability theory was his most influential contribution to mathematics. Originally applied to gambling, today it is extremely important in economics, especially in actuarial science. John Ross writes, "Probability theory and the discoveries following it changed the way we regard uncertainty, risk, decision-making, and an individual's and society's ability to influence the course of future events." However, it should be noted that Pascal and Fermat, though doing important early work in probability theory, did not develop the field very far. Christiaan Huygens, learning of the subject from the correspondence of Pascal and Fermat, wrote the first book on the subject. Later figures who continued the development of the theory include Abraham de Moivre and Pierre-Simon Laplace.

In literature, Pascal is regarded as one of the most important authors of the French Classical Period and is read today as one of the greatest masters of French prose. His use of satire and wit influenced later polemicists. The content of his literary work is best remembered for its strong opposition to the rationalism of René Descartes and simultaneous assertion that the main countervailing philosophy, empiricism, was also insufficient for determining major truths.

In France, prestigious annual awards, Blaise Pascal Chairs are given to outstanding international scientists to conduct their research in the Ile de France region.[33] One of the Universities of Clermont-Ferrand, France – Université Blaise Pascal – is named after him. The University of Waterloo, Ontario, Canada, holds an annual math contest named in his honour.

Pascalian theology has grown out of his perspective that we are, according to Wood, "born into a duplicitous world that shapes us into duplicitous subjects and so we find it easy to reject God continually and deceive ourselves about our own sinfulness".

Roberto Rossellini directed a filmed biopic (entitled *Blaise Pascal*) which originally aired on Italian television in 1971. Pascal was a subject of the first edition of the 1984 BBC Two documentary, *Sea of Faith*, presented by Don Cupitt.

REFERENCES

1. Boyer, Carl B., *A History of Mathematics*, Wiley, 1968.
2. Datta, B., and A.N. Singh, *History of Hindu Mathematics, Part I,* Motilal Banarsi, Das, 1935.
3. *World of Mathematics*, 2 vols., Gale, 2001.
4. "Brahmagupta," Vidyapatha, http://www.vidyapatha.com/scientists/brahmagupta.php February 17, 2006).
5. Noether, Gottfried. "Emmy Noether," in *Women of Mathematics, A Biobibliographic Sourcebook,* Louise Grinstein and Paul Campbell, Editors, Greenwood Press, 1987.
6. Olsen, Lynn M. *Women in Mathematics*, Massachusetts Institute of Technology, 1974.
7. Child, J.M., The Early Mathematical Manuscripts of Leibniz, Open Court Publishing, New York, 1920
8. Hofmann, J.E., Leibniz in Paris, Cambridge University Press, 1974
9. Parkinson, G.H.R. Leibniz, Logical Papers, Clarendon Press, Oxford 1966
10. Benson Mates, The Philosophy of Leibniz, New York: Oxford University Press, 1986
11. R. S. Woolhouse, (ed.), Gottfried Wilhelm Leibniz: Critical Assessments, Volumes I - IV, Routledge Publishing Co., 1993
12. Usha G. (2014). Teaching of Mathematics. Madurai: Shanlax Publications.
13. Kulbir Singh Sidhu, The Teaching of Mathematics, New Delhi: Sterling Publishers, 2008.
14. Papola, C. (2006). Teaching of Mathematics. New Delhi: Anmol Publications.
15. Unnati Bishnoi, (2012). Teaching of Mathematics, Meerut: Lall Book Depot.
16. Kulshreshtha, A. K. (2008). Teaching of Mathematics. Meerut: R.Lall Books Depot.
17. Sidhu, K. S. (2006). The teaching of Mathematics. New Delhi: Sterling Publishers private ltd.
18. http://www.ideafinder.com/history/inventors/lovelace.htm
19. “Ada Byron, Lady Lovelace.” Biographies of Famous Women Mathematicians. Retrieved from http://www.agnesscott.edu/lriddle/WOMEN/love.htm
20. Hennell, Michael. 1958. John Venn and the Clapham Sect. London: Lutterworth Press.
21. Venn, John. 1962. The logic of chance. New York: Chelsea Pub. Co.
22. Venn, John. 1973. The principles of inductive logic. New York: Chelsea Pub. Co. ISBN 0828402655 ISBN 9780828402651
23. http://www.scienceworld.wolfram.com/biography/Brahmagupta.html.
24. http://www.thefamouspeople.com/profiles/aryabhata-5427.php
25. http://www.famous-mathematicians.com/aryabhata/

26. http://www.famous-mathematicians.com/brahmagupta/
27. http://www.encyclopedia.com/topic/Brahmagupta.aspx
28. http://www.mapsofindia.com/who-is-who/science-technology/bhaskara-ii.html
29. http://www.newworldencyclopedia.org/entry/Bh%C4%81skara_II
30. http://veda.wikidot.com/bhaskaracharya
31. http://www.freepressjournal.in/bhaskaracharya-the-great-astronomer-and-mathematician/642379